A GEOMETRIC INTRODUCTION TO TOPOLOGY

C.T.C. WALL
University of Liverpool

DOVER PUBLICATIONS, INC.
New York

Copyright

Published in Canada by General Publishing Company, Ltd., 30 Lesmill Road, Don Mills, Toronto, Ontario.
Published in the United Kingdom by Constable and Company, Ltd., 3 The Lanchesters, 162–164 Fulham Palace Road, London W6 9ER.

Bibliographical Note

This Dover edition, first published in 1993, is an unabridged and unaltered republication of the work first published by the Addison-Wesley Publishing Company, Reading, Massachusetts, in 1972.

Library of Congress Cataloging-in-Publication Data

Wall, C. T. C. (Charles Terence Clegg)
A geometric introduction to topology / C.T.C. Wall.
p. cm.
Originally published: Reading, Mass. : Addison-Wesley, 1972.
Includes indexes.
ISBN 0-486-67850-4 (pbk.)
1. Algebraic topology. I. Title.
QA612.W3 1993
514′.2—dc20 93-37834
CIP

Manufactured in the United States of America
Dover Publications, Inc., 31 East 2nd Street, Mineola, N.Y. 11501

Preface

This book is intended to provide a first course in algebraic topology which really is suitable for teaching to undergraduates. An attempt is made to reduce necessary preliminaries to a minimum; even so, the course is unlikely to be suitable before the third year.

I have made a deliberate break from many of the existing treatments of thc subject. There are two main aspects of this. One is that I do not presuppose a course in general topology, but work solely with point-sets in Euclidean space. This has the advantage of presenting the student with illustrative examples, without an overwhelming number of definitions. Also, many of the particular properties of point-sets which are used (metrizability; in particular, normality and the Hausdorff property) do not then necessitate separate discussions. The second break from tradition is in avoiding the use of simplexes. I am convinced that the ordinary definition of homology of a simplicial complex, necessitating a lengthy development of the subject before topological invariance can be proved, gives a poor introductory course. I use only concepts whose topological invariance is self-evident; effectively the Čech definition of H^0 and H^1, and the singular definition of H_0.

The break from simplicial complexes permits us to return to the older traditions of studying general point-sets. The approach is twofold: we first introduce a geometrical idea and then, using algebra, built it up into a machine. At each stage functorial properties are emphasized, though categories and functors are not formally defined in this book.

The high point of this text is the proof of the Alexander duality theorem in the plane, which includes the Jordan curve theorem as a special case. Not all of Part I is necessary for this. In particular, though we use the results of Chapter 7, the proofs (which are more difficult than most in this book) may be omitted on a first reading. The later chapters are mostly devoted to emphasizing the relation of topological ideas to other branches of pure mathematics.

I have kept the book as short as possible, consistent with the aims of providing a full and not too rapid exposition, and of obtaining at least one really substantial result. For this reason, the fundamental group has been entirely excluded. It would be an appropriate subject for a companion volume, but is so large a topic that it needs a full book to itself, to give students a real chance of learning anything worth while about it.

At the end of each chapter there is a brief section to indicate how the material in the chapter can be further developed, with references for such developments. There is also a selection of exercises and problems. Those marked with an asterisk depend on ideas not covered in this book.

This book grew out of courses given by the author at Cambridge and Liverpool in 1963–67. I am indebted to Frank Adams for comments which substantially improved the presentation of Parts II and III.

Liverpool, England C.T.C.W.
January 1972

CONTENTS

PART 2 THE DUALITY THEOREM

PART 3 FURTHER RESULTS IN THE TOPOLOGY OF PLANE SETS

PART 0
PRELIMINARIES

0
NOTATIONS AND PREREQUISITES

NUMBERS

We assume familiarity with standard properties of the real numbers, $\mathbf{R}$; in particular, the fact that every bounded subset X of $\mathbf{R}$ has a least upper bound, or supremum, sup X and a greatest lower bound, or infimum, inf X. Euclidean space $\mathbf{R}^n$ has as its points x the sequences of n real numbers: $x = (x_1, \ldots, x_n)$, for example, the origin $0 = (0, \ldots, 0)$. The x_i are called the *coordinates* of the point x. The distance between two such points is defined by the Pythagorean formula

$$d(x, x') = \sqrt{(x_1 - x'_1)^2 + \cdots + (x_n - x'_n)^2}.$$

This satisfies the "triangle inequality"

$$d(x, x') + d(x', x'') \geqslant d(x, x'').$$

The complex numbers $\mathbf{C}$ can be thought of as $\mathbf{R}^2$ with extra structure given by multiplication:

$$(x_1, x_2)\cdot(y_1, y_2) = (x_1 y_1 - x_2 y_2, x_1 y_2 + x_2 y_1).$$

We usually write i for $(0, 1)$, and $x_1 + ix_2$ for (x_1, x_2), and often denote complex numbers by z. Some standard functions of $z = x_1 + ix_2$ are

$$\text{the real part} \quad \operatorname{Re} z = x_1,$$

$$\text{the imaginary part} \quad \operatorname{Im} z = x_2,$$

$$\text{the modulus} \quad |z| = \sqrt{x_1^2 + x_2^2} = d(z, 0), \quad \text{and}$$

$$\text{the argument} \quad \arg z = \theta, \quad \text{characterized by}$$

$$0 \leqslant \theta < 2\pi, \quad \sin\theta = x_2/|z|, \quad \cos\theta = x_1/|z|.$$

The set of all integers (whole numbers)—positive, negative or zero—is denoted by $\mathbf{Z}$.

SETS

We usually denote a set by a single letter, for example X. For members x of X, write $x \in X$. If X is defined as the set of "things" x for which some property $P(x)$ is true, we write

$$X = \{x : P(x)\}.$$

We adopt customary set-theoretic notation (which we suppose familiar), namely

$X \cup Y = \{x : x \in X \text{ or } x \in Y \text{ or both}\}$, the *union* of X and Y,

$X \cap Y = \{x : x \in X \text{ and } x \in Y\}$, the *intersection* of X and Y,

$X - Y = \{x \in X : x \text{ does not belong to } Y\}$, the *complement* of Y in X, and

$Y \subset X$ if all members of Y belong to X, i.e. Y is contained (or included) in X.

We call X and Y *disjoint* if their intersection is the *empty set* $\varnothing$, that is, has no members. When X is understood, we will refer to $X - Y$ simply as the complement of Y (for subsets $Y \subset X$).

We have standard notations for some particular sets. For intervals of real numbers, if $a < b$, we write

$$[a, b] = \{x \in \mathbf{R} : a \leqslant x \leqslant b\},$$
$$]a, b[= \{x \in \mathbf{R} : a < x < b\},$$
$$[a, b[= \{x \in \mathbf{R} : a \leqslant x < b\}, \text{ etc.}$$

In particular, write $I = [0, 1]$ for the standard unit interval,

$$\mathbf{R}^* = \mathbf{R} - \{0\} = \{x \in \mathbf{R} : x \neq 0\},$$
$$\mathbf{R}_+ = [0, \infty[\quad = \{x \in \mathbf{R} : x \geqslant 0\},$$
$$\mathbf{R}_+^* =]0, \infty[\quad = \mathbf{R}^* \cap \mathbf{R}_+.$$

In more dimensions, we have

$$D^n = \{x \in \mathbf{R}^n : d(x, 0) \leqslant 1\}, \quad \text{the unit disk;}$$
$$S^{n-1} = \{x \in \mathbf{R}^n : d(x, 0) = 1\}, \quad \text{the unit sphere,}$$

and, for any $x \in \mathbf{R}^n$, $r \in \mathbf{R}_+^*$,

$$U(x, r) = \{y \in \mathbf{R}^n : d(x, y) < r\}.$$

In general, a subset of $\mathbf{R}^n$ is called *bounded* if it is contained in some $U(x, r)$.

Finally, for any two sets X and Y, we have the (cartesian) product

$$X \times Y = \{(x, y): x \in X, y \in Y\}.$$

Do not confuse this notation with $\{x, y\}$, which denotes the set whose (two) elements are x and y.

MAPS

Given two sets X and Y, a map f from X to Y associates to each element x of X a well-determined $f(x) \in Y$. We write $f: X \to Y$ to express that f is a map from X to Y. We will not insist on any particular logical scheme, except to emphasize that if Y is a subset of Z, we do not regard f as a map from X to Z (though of course it defines one): the concept of map contains the sets X and Y as part of the structure. X is called the domain of f, and Y is the codomain. If f is defined by some formula expressing $f(x)$ in terms of x, we may write

$$x \mapsto f(x):$$

$f(x)$ is called the *image* of x under f.

If $f: X \to Y$ and $g: Y \to Z$, the *composite map*

$$g \circ f \text{ (or simply } gf): X \to Z$$

is defined by

$$(g \circ f)(x) = g(f(x)).$$

If $A \subset X$ is a subset, we have an *inclusion map* $i: A \to X$ defined by $i(a) = a$ for all $a \in A$. The composite of i with $f: X \to Y$ is denoted by

$$f \mid A: A \to Y,$$

and called the *restriction* of f to A; we say that f *extends* $f \mid A$. A rather trivial example of inclusion map is the *identity map* $1_X: X \to X$; when no confusion is likely, we simply denote this by 1. Another rather simple example of $f: X \to Y$ is obtained by choosing $y_0 \in Y$ and defining $f(x) = y_0$ for all $x \in X$. Such a map is called a *constant map*; it can also be defined as the composite of the (unique) map $X \to \{y_0\}$ and the inclusion $\{y_0\} \to Y$. We write (for any X, Y)

$$p_1: X \times Y \to X, \qquad p_2: X \times Y \to Y$$

for the projection maps defined by $p_1(x, y) = x$, $p_2(x, y) = y$.

A map $f: X \to Y$ is *injective* if different elements of X have different images, i.e. if

$$f(x) = f(x') \quad \text{implies} \quad x = x'.$$

It is *surjective* if for each $y \in Y$ there is an $x \in X$ with $f(x) = y$; a map which is both injective and surjective is said to be *bijective.*

If $f: X \to Y$ and $g: Y \to X$ with $g \circ f = 1_X$, then f is injective and g is surjective. The so-called Axiom of Choice states that, conversely, if g is surjective, there exists a map f with $g \circ f = 1_X$. Also g is bijective if and only if f is; this is equivalent to $f \circ g = 1_Y$. Notice that if $g \circ f$ is surjective, then so is g; if it is injective, then so is f.

A diagram of maps between various sets is said to *commute* or be *commutative* if all the composite maps in the diagram between any given pair of sets are the same—e.g. for the diagram

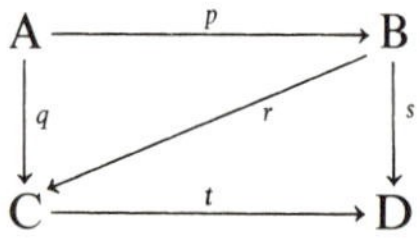

this means $r \circ p = q$, $t \circ r = s$, and (hence) $s \circ p = t \circ q$.

To conclude, here are two notations which are not (as the above all are) quite standard. Given sets X and Y, we write Map (X, Y) for the set of *all* maps $X \to Y$ (it is an axiom of set theory that this *is* a set). Finally, given a map $f: X \to Y$ and a subset B of Y we define

$$f^{\dashv}(B) = \{x \in X : f(x) \in B\}.$$

It is common to write $f^{-1}(B)$ for this, but this notation lends itself to confusion: ours is taken from a suggestion of Porteous ["Topological Geometry" (North-Holland)]. If A is a subset of X, we write, as is usual,

$$f(A) = \{f(a) : a \in A\}.$$

EQUIVALENCE RELATIONS

We frequently wish to describe a partition of a set X into subsets $\{X_\alpha\}$ of which no two have a common member, i.e. disjoint subsets. Call two elements x, x' of X *equivalent* if they belong to the same subset X_α, and write $x \sim x'$ for this. Then $\sim$ has three properties:

Reflexive $x \sim x$ (this means that each $x \in X$ belongs to some X_α).

Symmetric $x \sim y$ implies $y \sim x$.

Transitive $x \sim y$ and $y \sim z$ imply $x \sim z$.

Conversely, if $\sim$ has these properties, it is called an *equivalence relation*, and we obtain as above a partition of X into disjoint subsets, called *equivalence classes.*

1
SPACES AND CONTINUOUS MAPS

INTRODUCTION

In this chapter we will give the analytic topology necessary for the rest of the book, from as naive a viewpoint as possible. Thus we discuss exclusively subspaces of $\mathbf{R}^n$, though many readers will probably have met a more general definition of "topological space."

CONTINUITY

Topology is the geometrical study of continuity. This can, of course, be done axiomatically but the most important and interesting questions arise from the study of subsets of the euclidean spaces $\mathbf{R}^n$. We will write, briefly, *space* to denote a subset of some euclidean space.

If X and Y are spaces and $f: X \to Y$ a function, then we recall from analysis that f is said to be continuous at a point $x \in X$ if given any positive real number ε, we can choose a positive real number δ such that if $x' \in X$ and $d(x, x') < \delta$, then $d(f(x), f(x')) < \varepsilon$. We call f a continuous map if it is continuous at each point of X.

It is somewhat cumbersome to use this definition in practice, and it is usually more convenient to argue using some of the properties of continuity which we now give. The first follows at once from the definition.

C1 *Let $f: X \to Y$ be continuous: let $X' \subset X$ and $Y' \subset Y$ satisfy $f(X') \subset Y'$; denote the restriction of f by $f': X' \to Y'$. Then f' is continuous.* ∎

The next property is more important.

C2 *If $f: X \to Y$ and $g: Y \to Z$ are continuous, so is $g \circ f: X \to Z$.*

Proof. Let $x \in X$. We will show that $g \circ f$ is continuous at x. Let $\varepsilon > 0$.

Since g is continuous at $f(x)$, we can find $\eta > 0$ such that for $y \in Y$ and $d(y, f(x)) < \eta$ we have

$$d(g(y), g(f(x))) < \varepsilon.$$

Now since f is continuous at x, we can find $\delta > 0$ such that for $x \in X$ and $d(x', x) < \delta$ we have

$$d(f(x'), f(x)) < \eta$$

and therefore, by the above (putting $y = f(x')$),

$$d(g(f(x')), g(f(x))) < \varepsilon. \blacksquare$$

C3 *If $X \subset Y$, the inclusion map $X \to Y$ is continuous.*

This is obvious; take $\delta = \varepsilon$. ■

It follows that the codomain is not relevant to the problem whether a map is continuous or not: if $f: X \to Y'$, and $Y' \subset Y$ (with inclusion i) then by C1, $i \circ f$ continuous implies f continuous, and by C2 and C3, f continuous implies $i \circ f$ continuous. So if $Y \subset \mathbf{R}^m$, it is enough to find out whether f defines a continuous map from X to $\mathbf{R}^m$.

Now let $f: X \to \mathbf{R}^m$. Then f defines component maps $f_i: X \to \mathbf{R}$ $(1 \leqslant i \leqslant m)$ by

$$f(x) = (f_1(x), \ldots, f_m(x)) \qquad (x \in X).$$

C4 *For $f: X \to \mathbf{R}^m$, f is continuous if and only if the component maps $f_i: X \to \mathbf{R}$ are continuous.*

Proof. If f is continuous, then given $x \in X$ and $\varepsilon > 0$, choose δ as in the definition. Now

$$d(f(x), f(x')) = \sqrt{\sum_{i=1}^{m} d(f_i(x), f_i(x'))^2}$$
$$\geqslant d(f_i(x), f_i(x')),$$

so if $d(x, x') < \delta$, then

$$d(f_i(x), f_i(x')) < \varepsilon.$$

Conversely, if each f_i is continuous, choose $\delta_i > 0$ so that if $d(x, x') < \delta_i$, then

$$d(f_i(x), f_i(x')) < \varepsilon/\sqrt{n}.$$

Now if $d(x, x') < \delta = \min(\delta_i)$, it follows that

$$d(f(x), f(x')) < \varepsilon. \blacksquare$$

We shall want, for examples, the functions of elementary analysis.

C5 i) *Linear and constant maps* $\mathbf{R}^m \to \mathbf{R}$ *are continuous.*
ii) *Multiplication* $\mathbf{R}^2 \to \mathbf{R}$ *is continuous.*
iii) *Positive powers of x give continuous maps* $x \to x^\alpha$, $\mathbf{R}_+ \to \mathbf{R}(\alpha > 0)$.
iv) *The reciprocal is continuous,* $\mathbf{R}^* \to \mathbf{R}$.
v) *The exponential, sine and cosine functions are continuous,* $\mathbf{R} \to \mathbf{R}$.
vi) *The natural logarithm function is continuous,* $\mathbf{R}_+^* \to \mathbf{R}$.

The reader is probably familiar with these examples: we shall not give the proofs. ■

Perhaps more interesting is to see how C1 through C5 imply the continuity of other functions constructed using these. For example, if $f, g: X \to \mathbf{R}$ are continuous, then $f + g$, fg, and (if g does not vanish on X) g^{-1} are also continuous maps $X \to \mathbf{R}$. For by C4, the map $X \to \mathbf{R}^2$ given by $x \to (f(x), g(x))$ is continuous, by C5(i) and (ii), addition and multiplication are continuous maps $\mathbf{R}^2 \to \mathbf{R}$, and by C2 the composite maps $f + g$ and $fg: X \to \mathbf{R}$ are still continuous. If g is never zero on X, by C1 it gives a continuous map $X \to \mathbf{R}^*$ and by C5(iv) the reciprocal $\mathbf{R}^* \to \mathbf{R}$ is continuous, so by C2 the composite of these two, $g^{-1}: X \to \mathbf{R}$ is continuous.

It now follows by induction that if $f_1, \ldots, f_n$ are continuous maps $X \to \mathbf{R}$, so is any polynomial in the f_i. For example, if $X \subset \mathbf{R}^n$ and $x_1, \ldots, x_n$ are coordinate functions, we can use any polynomial in the x_i. If we have m such polynomials, they define a continuous map $X \to \mathbf{R}^m$; if the image lies in $Y \subset \mathbf{R}^m$, then by C1 we have a continuous map $X \to Y$. Similarly, but with a little more care, the continuity of more complicated examples can be verified. The reader should now be able to check the continuity of the maps given in examples below.

HOMEOMORPHISM

We can now make rather more precise the statement that topology is the study of continuity. Suppose we are given two spaces X, X' and continuous maps $f: X \to X'$, $f': X' \to X$ which are inverse to each other, i.e., such that $f'(f(x)) = x$ for each $x \in X$ and $f(f'(y)) = y$ for each $y \in X'$. Then f and f' are called *homeomorphisms* and X and X' are said to be homeomorphic or topologically equivalent. If $f: X \to Y$ has image X', and restricts to a homeomorphism $X \to X'$, f is called an *embedding.* (*Warning:* Such an f is called a homeomorphism in old-fashioned textbooks.)

To the topologist, homeomorphic spaces are indistinguishable: they have all the same topological properties, and a single word is often used—for example, any space homeomorphic to $[0, 1]$ is called an *arc.* For another example, a topologist cannot tell the difference between a circle and a square; and both are called *simple closed curves,* or *Jordan curves.*

Example (Fig. 1.1)

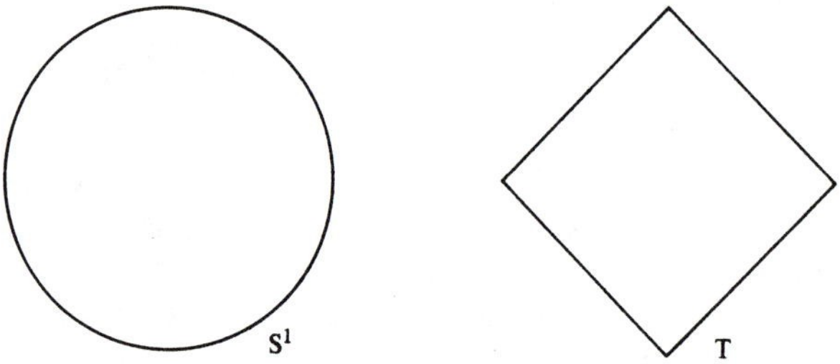

$S^1 = \{(x_1, x_2) : x_1^2 + x_2^2 = 1\}$

$T = \{(x_1, x_2) : |x_1| + |x_2| = 1\}$

Fig. 1.1

Inverse homeomorphisms $f: S^1 \to T$ and $f': T \to S^1$ are given by

$$f(x_1, x_2) = \left(\frac{x_1}{|x_1| + |x_2|}, \frac{x_2}{|x_1| + |x_2|}\right),$$

$$f'(x_1, x_2) = \left(\frac{x_1}{\sqrt{x_1^2 + x_2^2}}, \frac{x_2}{\sqrt{x_1^2 + x_2^2}}\right). \blacksquare$$

Example. The three spaces in Fig. 1.2 are homeomorphic. For if we define $h: X_1 \to X_2$ by

$$h(x_1, x_2) = \left(\frac{x_1}{\sqrt{x_1^2 + x_2^2}}, \frac{x_2}{\sqrt{x_1^2 + x_2^2}}, \tfrac{1}{2}\log(x_1^2 + x_2^2)\right),$$

then h is continuous, and so is h' below, which is inverse to h:

$$h'(x_1, x_2, x_3) = (x_1 e^{x_3}, x_2 e^{x_3}).$$

The punctured plane, X_1 | A circular cylinder, X_2 | A hyperboloid of one sheet, X_3

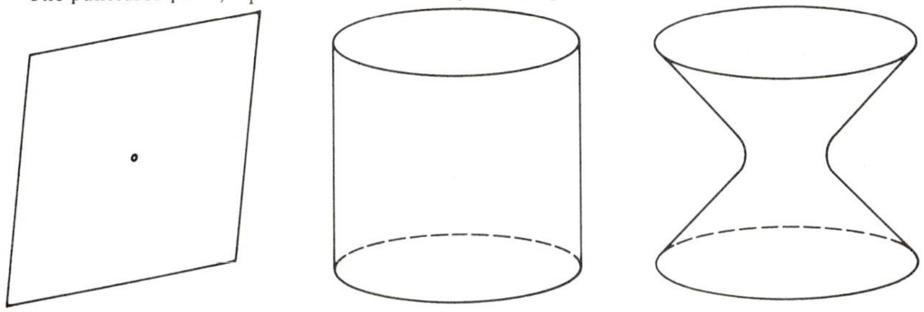

$X_1 = \{(x_1, x_2) : (x_1, x_2) \neq (0,0)\}$ | $X_2 = \{(x_1, x_2, x_3) : x_1^2 + x_2^2 = 1\}$ | $X_3 = \{(x_1, x_2, x_3) : x_1^2 + x_2^2 - x_3^2 = 1\}$

Fig. 1.2

Similarly we have $k: X_2 \to X_3$ with inverse k':

$$k(x_1, x_2, x_3) = (x_1\sqrt{1 + x_3^2}, x_2\sqrt{1 + x_3^2}, x_3),$$
$$k'(x_1, x_2, x_3) = \big(x_1(1 + x_3^2)^{-1/2}, x_2(1 + x_3^2)^{-1/2}, x_3\big). \blacksquare$$

However, a topologist draws a very sharp distinction between the circle S^1 described above and the space

$$D^2 = \{(x_1, x_2): x_1^2 + x_2^2 \leqslant 1\},$$

which we call a *ball* (or disk, or cell). The two spaces do not look alike (see Fig. 1.3). We will show in Chapter 6 that they are not homeomorphic.

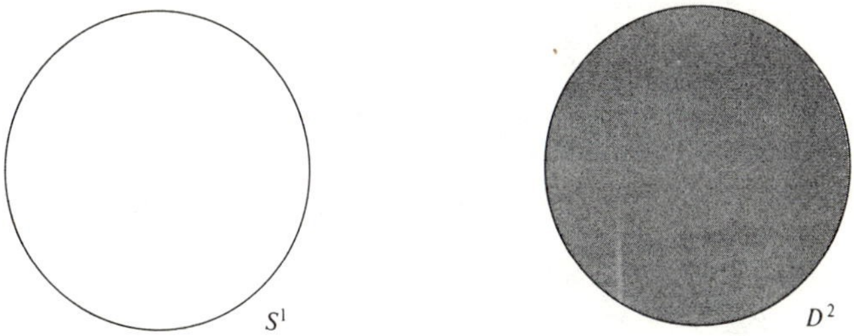

Fig. 1.3

For further examples, the concept of product will be useful. If $X \subset \mathbf{R}^m$ and $Y \subset \mathbf{R}^n$, we define $X \times Y$ to be the set of points $(x_1, \ldots, x_{m+n})$ in $\mathbf{R}^{m+n}$ such that $(x_1, \ldots, x_m)$ defines a point of X, and $(x_{m+1}, \ldots, x_{m+n})$ a point of Y.

The topological nature of this construction is left as an exercise to the reader. Observe that the cylinder in the example above is the product $S^1 \times \mathbf{R}$. The next example shows that unique factorization does not hold for this kind of product.

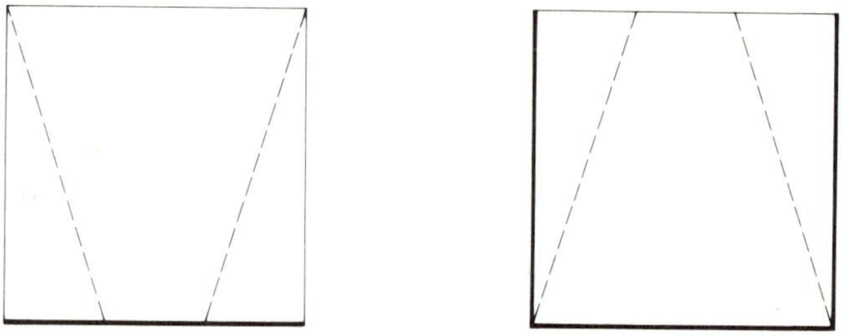

Fig. 1.4.

Example The spaces $]0, 1[\times [0, 1[$, and $[0, 1] \times [0, 1[$ are homeomorphic. Divide each square in Fig. 1.4 into three regions as indicated by the dashed lines. Now define f on these respective regions by

$$f(x_1, x_2) = \begin{cases} \left(\dfrac{x_2}{3}, 1 - 3x_1\right) & \text{if } 3x_1 + x_2 \leqslant 1, \\ \left(x_1 + (1 - 2x_1)\left(\dfrac{2x_2 - 1}{2x_2 + 1}\right), x_2\right) & \text{if } 1 - x_2 \leqslant 3x_1 \leqslant 2 + x_2, \\ \left(1 - \dfrac{x_2}{3}, 3x_1 - 2\right) & \text{if } 3x_1 - x_2 \geqslant 2. \end{cases}$$

We leave it to the reader to check that this is a homeomorphism.

Example The group U_2 of 2×2 unitary (complex) matrices is homeomorphic to $S^3 \times S^1$. We define $f: S^3 \times S^1 \to U_2$ by

$$f\{(x_0, x_1, x_2, x_3), (y_0, y_1)\} = \begin{pmatrix} x_0 + ix_1 & x_2 + ix_3 \\ -(y_0 + iy_1)(x_2 - ix_3) & (y_0 + iy_1)(x_0 - ix_1) \end{pmatrix},$$

and $g: U_2 \to S^3 \times S^1$ by

$$g\begin{pmatrix} z_0 & z_1 \\ z_2 & z_3 \end{pmatrix} = \{(x_0, x_1, x_2, x_3), (y_0, y_1)\},$$

with

$$x_0 + ix_1 = z_0,$$
$$x_2 + ix_3 = z_1,$$
$$y_0 + iy_1 = z_0 z_3 - z_1 z_2.$$

These maps are clearly continuous; we leave to the reader to verify that their ranges are as stated, and that they are inverse to each other. ■

NEIGHBORHOODS, OPEN AND CLOSED SETS

We now reformulate the definition of continuity. Let X be a space, $x \in X$. Then the (open) δ-neighborhood of $x \in X$ is defined as

$$U_X(x, \delta) = \{x' \in X : d(x, x') < \delta\}.$$

We omit the X if there is no risk of confusion. Any subset of X containing $U(x, \delta)$ for some $\delta > 0$ is called a neighborhood of x. A subset of X is called *open* (in X) if it is a neighborhood of each of its points.

1.1 Theorem *Let X, Y be spaces, $f: X \to Y$. The following are equivalent:*
1) *f is continuous at $x \in X$,*
2) *for any neighborhood N in Y of $f(x)$, $f^{-1}(N)$ is a neighborhood of x.*

The following are also equivalent:

3) *f is continuous*
4) *for any V open in Y, $f^{-1}(V)$ is open in X.*

Proof In the notation just introduced, the definition of (1) becomes: given $\varepsilon > 0$, we can choose $\delta > 0$ such that

$$f\big(U(x, \delta)\big) \subset U\big(f(x), \varepsilon\big),$$

or equivalently,

$$U(x, \delta) \subset f^{-1}\big(U(f(x), \varepsilon\big).$$

The existence of δ means that $f^{-1}\big(U(f(x), \varepsilon\big)$ is a neighborhood of x. Thus (2) implies (1). Conversely, if we assume this, then since any neighborhood N of $f(x)$ contains some $U\big(f(x), \varepsilon\big)$, it follows that

$$f^{-1}(N) \supset f^{-1}\big(U(f(x), \varepsilon\big),$$

and so is a neighborhood of x. This establishes the equivalence of (1) and (2).

Now let us assume f continuous, and let V be open in Y. To establish (4), we must show that $f^{-1}(V)$ is a neighborhood of each of its points. If $x \in f^{-1}(V)$, that is $f(x) \in V$, we know that V is a neighborhood of $f(x)$. Since (2) holds, $f^{-1}(V)$ is a neighborhood of x.

To show that (4) implies (3) we need a lemma.

1.2 Lemma *Let X be a space, $x \in X$, and $\delta > 0$. Then $U(x, \delta)$ is an open subset of X.*

Proof. We must show that $U(x, \delta)$ is a neighborhood of each of its points. Let $y \in U(x, \delta)$. Then $d(x, y) < \delta$. Let $\varepsilon = \delta - d(x, y)$. Then $U(y, \varepsilon) \subset U(x, \delta)$, since if $z \in U(y, \varepsilon)$, we have, by the triangle inequality,

$$\begin{aligned} d(x, z) &\leqslant d(x, y) + d(y, z) \\ &< d(x, y) + \varepsilon = \delta. \end{aligned}$$

So $U(x, \delta)$ is a neighborhood of y, as claimed. ■

We return to the proof of the theorem. We assume (4), and have to show that for every x in X and $\varepsilon > 0$,

$$U = f^{-1}(U(f(x), \varepsilon))$$

is a neighborhood of x. But $U(f(x), \varepsilon)$ is open, hence by (4), so is U; now since U is open and $x \in U$, U is a neighborhood of x. ■

The idea of open set is commonly used as the foundation stone of topology. We now deduce some properties of open sets according to the above definition, which are used as the axiomatic basis of general topology.

1.3 Proposition *For any space X, ϕ and X are open in X; the intersection of two open sets is open, and the union of any family of open sets is open.*

Proof. All these are immediate consequences of the definitions, except for the assertion about the intersection of open sets. Let U, V be open in X; let $x \in U \cap V$. Then U and V are both neighborhoods of x; let $U(x, \delta) \subset U$ and $U(x, \varepsilon) \subset V$. Then

$$U(x, \min(\delta, \varepsilon)) \subset U \cap V,$$

so $U \cap V$ is a neighborhood of x. Since this is true for each $x \in U \cap V$, it follows that $U \cap V$ is open. ■

Example Let $X \subset \mathbf{R}$ be the set $\mathbf{Z}$ of integers. For each integer n, $n = U_{\mathbf{Z}}(n, \delta)$ for any $\delta < 1$. Thus $\{n\}$ is an open subset of $\mathbf{Z}$. Hence any subset of $\mathbf{Z}$ is open. In general, a space is called *discrete* if all its subsets are open.

Since any union of open sets is open, we can consider for any $Y \subset X$ the union of the open subsets of X which are contained in Y. This is the largest open subset of X contained in Y, and is called the *interior* of Y, and written $\mathrm{Int}(Y)$ or $\mathrm{Int}_X(Y)$. Thus Y is open in X if and only if it coincides with $\mathrm{Int}_X Y$. The set of points of X which are interior neither to Y nor to $X - Y$ is called the frontier (in X) of Y, written $\mathrm{Fr}_X(Y)$.

A subset F of X is *closed* in X if $X - F$ is open. Any property of closed sets can thus be stated in terms of open sets, or vice versa. For example, given any subset Y of X, there is a smallest closed subset of X containing Y. This is called the *closure* of Y in X, and written $\mathrm{Cl}_X(Y)$. It is the union of the interior and the frontier of Y in X. Y is closed in X only if $Y = \mathrm{Cl}_X(Y)$.

For any space X, nonempty subset Y, and $x \in X$, write

$$d(x, Y) = \inf\{d(x, y) : y \in Y\}.$$

1.4 Lemma *For $x, x' \in X$, $|d(x, Y) - d(x', Y)| \leqslant d(x, x')$. Thus $d(x, Y)$ is a continuous function of x.*

Proof. For any $\varepsilon > 0$ choose $y \in Y$ with

$$d(x', y) < d(x', Y) + \varepsilon.$$

Then

$$d(x, Y) \leqslant d(x, y) \leqslant d(x, x') + d(x', y)$$
$$< d(x, x') + d(x', Y) + \varepsilon.$$

This holds for all $\varepsilon > 0$, so

$$d(x, Y) \leqslant d(x, x') + d(x', Y).$$

Similarly,

$$d(x', Y) \leqslant d(x, x') + d(x, Y),$$

which proves the first assertion. The second follows from the definition of continuity (we can take $\delta = \varepsilon$). ■

We can now characterize the closure.

1.5 Lemma *Let X be a space, Y a nonempty subspace, and $x \in X$. The following are equivalent:*

i) $x \in \mathrm{Cl}_X(Y)$.

ii) *Any open set containing x intersects Y.*

iii) $d(x, Y) = 0$.

Proof. It is easier to see directly the equivalence of the opposite statements:
i)′ $x \notin \mathrm{Cl}_X(Y)$, that is, $x \in \mathrm{Int}_X(X - Y)$.
ii)′ For some open set U, $x \in U \subset X - Y$.
iii)′ $d(x, Y) > 0$.

For (i)′ and (ii)′ are equivalent by definition of "Int," and (ii)′ holds for some U if and only if for some $\delta > 0$,

$$U_X(x, \delta) \subset X - Y,$$

that is,

$$d(x, Y) \geqslant \delta. \blacksquare$$

Corollary *If Y is closed in X and $x \in X - Y$, then $d(x, Y) > 0$.* ■

The following is often useful in arguments using open sets.

1.6 Lemma *If X is a space, $Y \subset X$, then the open (closed) subsets of Y are those of the form $A \cap Y$ with A open (closed) in X.*

Proof. If A is closed in X and $y \in Y$, then

$$d(y, A \cap Y) = 0 \qquad \text{implies} \qquad d(y, A) = 0$$

so $y \in A$, that is, $y \in A \cap Y$. Thus $A \cap Y$ is closed in Y. Conversely, if B is closed in Y, put $A = \mathrm{Cl}_X(B)$. Then if $y \in A \cap Y$, we have $d(y, B) = 0$, so as B is closed in Y, $y \in B$. Thus $B = A \cap Y$. The result for open sets follows by taking complements. ■

Corollary *If also Y is open (closed) in X, the open (closed) subsets of Y are those which are also open (closed) in X.*

For the intersection of two open (closed) subsets of X is again open (closed). ■

We now give a criterion for continuity which will be used repeatedly in the rest of the book.

1.7 Theorem *Let X be a space, X_1 and X_2 closed subsets with $X = X_1 \cup X_2$. Let $f: X \to Y$ be a map whose restrictions $f_1: X_1 \to Y$ and $f_2: X_2 \to Y$ are continuous. Then f is continuous.*

Proof. We prove continuity at each $x \in X$. First let $x \in X_1 - X_2$. Since f_1 is continuous at x for any $\varepsilon > 0$, $f_1^{-1}(U(f(x), \varepsilon))$ is a neighborhood of x in X_1. Say it contains $U_{X_1}(x, \delta)$. By the corollary to Lemma 1.5, $d(x, X_2) > 0$. If $\delta_1 = \min(\delta, d(x, X_2))$, then

$$U_{X_1}(x, \delta) \supset U_X(x, \delta_1),$$

so we have a neighborhood of x in X. The case $x \in X_2 - X_1$ is similar.

Now let $x \in X_1 \cap X_2$. Then we may suppose that

$$f_1^{-1}(U(f(x), \varepsilon)) \supset U_{X_1}(x, \delta_1), \qquad f_2^{-1}(U(f(x), \varepsilon)) \supset U_{X_2}(x, \delta_2).$$

But then it follows that the union

$$f^{-1}(U(f(x), \varepsilon)) \supset U_{X_1}(x, \delta_1) \cup U_{X_2}(x, \delta_2) \supset U_X(x, \min(\delta_1, \delta_2)). \blacksquare$$

It follows by induction that a corresponding result holds if X is the union of a finite collection of closed subsets $X_i (1 \leqslant i \leqslant n)$.

COMPACTNESS

A space X is said to be *compact* if, given any collection $\{U_\alpha\}$ of open sets in X with $X = \bigcup U_\alpha$, we can find a finite subcollection $U_{\alpha_1}, \ldots, U_{\alpha_n}$ with $X = U_{\alpha_1} \cup U_{\alpha_2} \cup \cdots \cup U_{\alpha_n}$.

The above is the intrinsic definition, describing the most important property of compact spaces. However, there is also an easier characterization.

1.8 Theorem *A space $X \subset \mathbf{R}^n$ is compact if and only if it is closed and bounded.*

We shall give the proof in a sequence of lemmas.

1.9 Lemma *Let $X \subset Y$, with X compact. Then X is a closed subset of Y.*

Proof. Suppose $y \in Y$, $y \notin X$. Then for each $x \in X$, $d(x, y) > 0$. Now the sets

$$U_X(x, \tfrac{1}{2}d(x, y)) \qquad \text{for } x \in X$$

are open in X, and their union is X (since each $x \in X$ belongs to the disk with itself as center). Since X is compact, we can find a finite subcollection, corresponding say to $x_1, \ldots, x_n$, whose union is already X. Thus for any $x \in X$ there exists i $(1 \leqslant i \leqslant n)$ with

$$x \in U_X(x_i, \tfrac{1}{2}d(x_i, y)).$$

Then $d(x, x_i) < \frac{1}{2}d(x, y)$ and so, by the triangle inequality,

$$d(x, y) \geqslant \tfrac{1}{2}d(x_i, y) \geqslant \tfrac{1}{2} \min_{1 \leqslant i \leqslant n} d(x_i, y).$$

Now by Lemma 1.6, $y \notin \mathrm{Cl}_Y X$. Since this holds for all $y \notin X$, X is closed in Y. ■

Corollary *A subset A of a compact space X is closed in X if and only if it is compact.*

For the lemma shows that if A is compact, it is closed in X. Conversely, if A is closed, and $\{U_\alpha\}$ is a collection of open subsets of A with union A, then for each α, $A - U_\alpha$ is closed in A, and hence in X, so

$$X - (A - U_\alpha) = (X - A) \cup U_\alpha$$

is open in X. Next,

$$\bigcup\{(X - A) \cup U_\alpha\} = (X - A) \cup \bigcup U_\alpha = (X - A) \cup A = X.$$

Since X is compact, we can pick a finite subset $\alpha_1, \ldots, \alpha_n$ such that

$$X = \bigcup_{1 \leqslant i \leqslant n} \{(X - A) \cup U_{\alpha_i}\}.$$

But then

$$A = \bigcup_{1 \leqslant i \leqslant n} U_{\alpha_i},$$

showing that A is compact. ■

Proof of Theorem 1.8. First suppose X is compact. Then by Lemma 1.9 it is closed in $\mathbf{R}^n$. Now the open subsets $X \cap U(0, m)$, for $m > 0$, have union X. Hence we can find a finite collection with union X. Let them correspond to $m_1, \ldots, m_k$. But if $N = \max(m_1, \ldots, m_k)$, it then follows that $X \subset U(0, N)$, that is, that X is bounded.

Now suppose X is closed and bounded, say $X \subset U(0, N)$. Then X is contained in the cube C given by $|x_i| \leqslant N$ for each i. Moreover, X is a closed subset of C. If we can show that C is compact, the corollary above will then show that X is compact too. Thus the theorem follows from the next lemma.

1.10 Lemma (*The Heine-Borel theorem*). *A closed cube C in $\mathbf{R}^n$ is compact.*

Proof. We will suppose $\bigcup U_\alpha = C$, with U_α open in C, and that no finite subfamily has union C, and will obtain a contradiction.

Divide C into 2^n equal cubes by hyperplanes parallel to its faces. Some of these smaller cubes C' may have the property that we can find a finite set $\alpha_1, \ldots, \alpha_k$ with $C' \subset U_{\alpha_1} \cup \cdots \cup U_{\alpha_k}$. However, this cannot occur for all the C', else the union of the finite set of all these U_{α_i} would be C. Thus we can choose a C' without the property, call it C_1.

Now divide C_1 into 2^n equal cubes, and repeat the argument. We obtain inductively a sequence of cubes

$$C \supset C_1 \supset C_2 \supset \cdots \supset C_r \supset C_{r+1} \supset \cdots$$

such that, for each r, the length of an edge of C_r is 2^{-r} times that of C, and there is no finite set $\alpha_1, \ldots, \alpha_k$ with $C_r \subset U_{\alpha_1} \cup \cdots \cup U_{\alpha_k}$. We next observe that those cubes intersect in a single point. For if α_n, β_n are the least and greatest values respectively of the first coordinate on the cube C_n, then the sequence α_n is increasing and bounded, thus tends to a limit; since $\beta_{n+1} - \alpha_{n+1} = \frac{1}{2}(\beta_n - \alpha_n)$, the sequence β_n tends to the same limit ξ_1, say. Similarly we find unique values ξ_i for the other coordinates so that $P = (\xi_1, \ldots, \xi_n)$ lies in all the cubes C_r.

But the point P itself lies in some open set U_α. Since U_α is open, we can find $\varepsilon > 0$, so that $d(P, Q) < \varepsilon$ implies $Q \in U_\alpha$. Now if C has diameter δ, C_n has diameter $2^{-n}\delta$. Thus, if n is so large that $2^{-n}\delta < \varepsilon$, we find that for all $Q \in C_n$,

$$d(P, Q) \leqslant 2^{-n}\delta < \varepsilon$$

so that $Q \in U_\alpha$; thus $C_n \subset U_\alpha$. This contradicts the property which C_n was supposed to have by construction. The theorem is thus proved. ■

The above proof mentions one of the characteristic properties of compact spaces: that they are closed in any space which contains them. Another is their behavior under mappings.

1.11 Theorem *If X is a compact space, and $f: X \to Y$ a continuous surjective map, then Y is compact.*

Proof. Let U_α be a collection of open subsets of Y whose union is Y. Since f is continuous, the $f^{-1}(U_\alpha)$ are open subsets of X. Moreover, for any $x \in X$, $f(x) \in Y$ must be in some U_α, so $x \in f^{-1}(U_\alpha)$, so the union of these open sets is all of X. But X is compact, so we can choose

$$X = f^{-1}(U_{\alpha_1}) \cup \cdots \cup f^{-1}(U_{\alpha_n}).$$

But since f is surjective, $Y = U_{\alpha_1} \cup \cdots \cup U_{\alpha_n}$. This shows that Y is compact. ■

The following consequence of this result will be useful.

Corollary *If X is a compact space, and $f: X \to Y$ a continuous bijective map, then f is a homeomorphism.*

Proof. As f is bijective, it has an inverse f^{-1}. What we must show is that f^{-1} is continuous, i.e. (by Theorem 1.1) that for U open in X, $(f^{-1})^{-1} U = f(U)$ is open in Y. Substituting A for $X - U$, this is equivalent to showing that for A closed in X, $f(A)$ is closed in Y. But since X is compact and A closed in X, by the corollary to Lemma 1.9 A is compact; now by Theorem 1.11, $f(A)$ is compact, and by Lemma 1.9, it follows that $f(A)$ is closed in Y. ■

1.12 Theorem *Suppose that A, B are disjoint closed subsets of $\mathbf{R}^n$, with A compact. Then*

$$d(A, B) = \inf\{d(a, b): a \in A, b \in B\} > 0.$$

Proof. Consider the open subsets $\{U_A(a, \frac{1}{2}d(a, B)): a \in A\}$ of A, with union A. Since A is compact, we can find a finite set $a_1, \ldots, a_k \in A$ such that the corresponding open sets have union A. Now for each $a \in A$, we can find i $(1 \leqslant i \leqslant k)$ with $d(a, a_i) < \frac{1}{2}d(a_i, B)$ and so by Lemma 1.4,

$$d(a, B) \geqslant \tfrac{1}{2}d(a_i, B) \geqslant \tfrac{1}{2} \min_{1 \leqslant i \leqslant k} d(a_i, B) = \delta,$$

say. It follows that $d(A, B) \geqslant \delta > 0$. ■

FURTHER DEVELOPMENTS

Analytic topology is a branch of mathematics in its own right, and there is no dearth of suitable reading in this field. Introductory books that can be recommended are:

Bourbaki, N., *Topologie Générale*, Hermann, Paris (issued in several parts). English translation, *General Topology*, Addison-Wesley, Reading, Mass., 1966.

Hocking, J. G. and G. S. Young, *Topology*, Addison-Wesley, Reading, Mass., 1961. (Similar viewpoint to this book.)

Hu, S. T., *Elements of General Topology*, Holden-Day, San Francisco, 1964. (Useful for algebraic topology.)

Kelley, J. L., *General Topology*, Van Nostrand, Princeton, 1955.

The texts by Kelley and Bourbaki are the classics of the American and French schools.

EXERCISES AND PROBLEMS

1. Show that the mod function $z \to |z|$ gives a continuous map $\mathbf{C} \to \mathbf{R}$.
2. Verify (a) that the functions described in the examples are continuous, and (b) that they give bijections.
3. Show that orthogonal projection defines a homeomorphism from the paraboloid $z = x^2 + y^2$ in $\mathbf{R}^3$ to the plane $z = 0$.
4. a) Show that every finite space is discrete.
   b) Show that every discrete space has a finite or countably infinite number of points. [*Hint:* If it is bounded, it must be finite.]
5. a) Prove that the open intervals $]0, 1[$, $]0, \infty[$, $]-\infty, \infty[$ on $\mathbf{R}$ are all homeomorphic.
   b) Prove that no two of $]0, 1[$, $[0, 1[$, $[0, 1]$ are homeomorphic. [*Hint:* First show that for any interval J, a continuous injective map $J \to \mathbf{R}$ is monotone.]
   c) Construct a homeomorphism between the products
   $$[0, 1] \times [0, 1[\qquad \text{and} \qquad [0, 1[\times [0, 1[.$$
6. Show that for $n \geqslant 1$, $\mathbf{R}^n - \{0\}$ is homeomorphic to $S^{n-1} \times \mathbf{R}$.
7. Let N (the "North Pole") be the point of S^n with coordinates $(0, \ldots, 0, 1)$. Define a map $f: S^n - \{N\} \to \mathbf{R}^n$ by projecting from the point N to the plane $x_{n+1} = 0$. Express the coordinates of $f(P)$ in terms of those of P, and vice-versa. Deduce that f is a homeomorphism.
8. a) Show that the quadric in $\mathbf{R}^n$ defined by
   $$x_1^2 + \cdots + x_p^2 - x_{p+1}^2 - \cdots - x_{p+q}^2 = 1 \qquad (p + q \leqslant n)$$
   is homeomorphic to $S^{n-1} \times \mathbf{R}^{n-p}$.
   b) Show that any noncentral quadric in $\mathbf{R}^n$ is homeomorphic to $\mathbf{R}^{n-1}$ (cf. Exercise 3).
9. Let $f: \mathbf{R}^2 \to \mathbf{R}$ be continuous; write
   $$A = \{P \in \mathbf{R}^2 : f(P) < 0\} \qquad \text{and} \qquad B = \{P \in \mathbf{R}^2 : f(P) \leqslant 0\}.$$
   Show that
   $$\text{Cl}_{\mathbf{R}^2} A \subset B \qquad \text{and} \qquad A \subset \text{Int}_{\mathbf{R}^2} B.$$
   Give examples of functions f for which these inclusions are strict; i.e. equality does not hold.
10. Find two closed subsets A, B of $\mathbf{R}^2$ with $A \cap B = \varnothing$ but $d(A, B) = 0$.
11. Show that if A and B are compact, there exist $x \in A$, $y \in B$ with $d(x, y) = d(A, B)$. Is this also true if A is compact and B is closed?
12. If X and Y are nonempty, show that $X \times Y$ is compact if and only if both X and Y are.
13. Let $ad - bc > 0$. Show that there is just one way to express the matrix
   $$\begin{pmatrix} a & b \\ c & d \end{pmatrix}$$

as a product

$$\begin{pmatrix} \cos\theta & \sin\theta \\ -\sin\theta & \cos\theta \end{pmatrix} \begin{pmatrix} p & q \\ 0 & r \end{pmatrix}$$

with $p > 0$, $r > 0$. Deduce that the space $\{(a, b, c, d) \in \mathbf{R}^4 : ad > bc\}$ is homeomorphic to $S^1 \times \mathbf{R}^3$.

14. Show that if $X = X_1 \cup \cdots \cup X_n$, and each X_i is compact, then X is compact.

15. Show that for any nonempty spaces X, Y, Z, a map $X \to Y \times Z$ is continuous if and only if the component maps $X \to Y$ and $X \to Z$ are.

2
ABELIAN GROUPS

INTRODUCTION

Although one of the objectives of this book is to show how topological problems lead to algebraic ones, we will not need very much algebra to solve these. In fact, we only use abelian groups, and most of those will be free. Thus the algebra needed resembles vector space theory. In this chapter we summarise the properties to be used.

DEFINITIONS

An abelian group is a set A together with a mapping $A \times A \to A$, usually denoted by

$$(a, b) \mapsto a + b,$$

such that

G1 There is an element $0 \in A$ with $0 + a = a$ for all $a \in A$.

G2 For each $a \in A$, there exists $(-a) \in A$ with $(-a) + a = 0$.

G3 For all $a, b, c \in A$, $(a + b) + c = a + (b + c)$.

G4 For all $a, b \in A$, $a + b = b + a$.

Axioms G1 through G3 define the concept of group. The most important example, for us, will be the group **Z** of integers with the usual addition operation. Other examples are the additive groups **R** of real, or **C** of complex, numbers. Note also the trivial group $\{0\}$ with only one element. We usually write $a - b$ for $a + (-b)$, $a + b + c$ for $a + (b + c)$, and so on.

Another standard notation for the same concept is the multiplicative. Here the basic operation becomes

$$(a, b) \mapsto ab,$$

and the identities required by the axioms are

$$1)\quad 1a = a, \qquad 2)\quad a^{-1}a = 1,$$

$$3)\quad (ab)c = a(bc), \qquad 4)\quad ab = ba.$$

Examples of groups in this notation are the multiplicative group $\mathbf{R}^*$ of nonzero real (or $\mathbf{C}^*$ of nonzero complex) numbers. Note that zero must be excluded here, otherwise (2) would be violated. More important for us will be the group S^1 of complex numbers of modulus 1, again with the usual multiplication of complex numbers. We leave to the reader (Exercise 2) to check that our examples do satisfy the axioms.

Let A be an abelian group. A subset B of A is a *subgroup* if

S1 $0 \in B$.

S2 If $b \in B$, then $(-b) \in B$.

S3 If $a, b \in B$, then $(a + b) \in B$.

Note that S3 states that addition in A restricts to give a map $B \times B \to B$; S1 and S2 imply that this restriction satisfies G1 and G2. Since it also satisfies G3 and G4, B is a group. For example, $\mathbf{Z}$ is a subgroup of $\mathbf{R}$, and both are subgroups of $\mathbf{C}$; $\{0\}$ can be regarded as a subgroup of every group.

Given two groups A, X, a map $\phi\colon A \to X$ is a *homomorphism* if

H1 For all $a, b \in A$, $\phi(a + b) = \phi(a) + \phi(b)$.

We leave it as an exercise (Exercise 2) to show that this implies that $\phi(0) = 0$ and $\phi(-a) = -\phi(a)$ for all $a \in A$. Clearly if $\phi\colon A \to B$ and $\psi\colon B \to C$ are homomorphisms, then so is $\psi \circ \phi\colon A \to C$. We will write $\mathrm{Hom}(A, X)$ for the set of all homomorphisms $A \to X$.

A bijective homomorphism $\phi\colon A \to X$ is called an *isomorphism* (between A and X). Isomorphisms play the same role in group theory as do homeomorphisms in topology. If ϕ is an isomorphism, so is ϕ^{-1}, and any algebraic properties possessed by A (number of elements, of subgroups, distribution of elements in subgroups, etc.) are all possessed by X; A and X are not abstractly distinguishable. For example, any group isomorphic to $\mathbf{Z}$ is called an infinite cyclic group.

If $\phi\colon A \to X$ is a homomorphism, we define its *kernel* and *image* by

$$\operatorname{Ker}\phi = \phi^{-1}\{0\} = \{a \in A : \phi(a) = 0\},$$

$$\operatorname{Im}\phi = \phi(A) = \{\phi(a) : a \in A\}.$$

2.1 Lemma *If $\phi\colon A \to X$ is a homomorphism, then* Ker ϕ *is a subgroup of A and* Im ϕ *is a subgroup of X. ϕ is injective if and only if* Ker $\phi = \{0\}$, *and surjective if and only if* Im $\phi = X$.

We will prove the third assertion and leave the others as exercises (Exercise 3). Suppose ϕ injective. Then if $a \in \operatorname{Ker} \phi$,

$$\phi(a) = 0 = \phi(0),$$

so since ϕ is injective, $a = 0$. Thus $\operatorname{Ker} \phi = \{0\}$. Conversely, let $\operatorname{Ker} \phi = \{0\}$, and suppose $\phi(a) = \phi(b)$. Then

$$\begin{aligned}\phi(a - b) &= \phi(a + (-b)) = \phi(a) + \phi(-b) \\ &= \phi(a) - \phi(b) \\ &= \phi(a) - \phi(a) = 0.\end{aligned}$$

Since $\operatorname{Ker} \phi = \{0\}$, $a - b = 0$, that is, $a = b$. This proves that ϕ is injective. ■

If $\phi\colon A \to X$ and $\psi\colon A \to X$ are homomorphisms, and we define $\phi + \psi$ in the obvious way,

H2 $(\phi + \psi)(a) = \phi(a) + \psi(a)$,

then $\phi + \psi$ is a homomorphism, for

$$\begin{aligned}(\phi + \psi)(a + b) &= \phi(a + b) + \psi(a + b) \\ &= \phi(a) + \phi(b) + \psi(a) + \psi(b) \\ &= \phi(a) + \psi(a) + \phi(b) + \psi(b) \\ &= (\phi + \psi)(a) + (\phi + \psi)(b),\end{aligned}$$

where we use in turn H2, H1, G4, and H2 again.

Thus the set $\operatorname{Hom}(A, X)$ is endowed with an addition map. It is not difficult to verify that this map satisfies G1 through G4, so that $\operatorname{Hom}(A, X)$ is itself an abelian group; for example, we have the zero homomorphism $0 \in \operatorname{Hom}(A, X)$ with $0(a) = 0$ for all $a \in A$.

DIRECT SUMS

Let A, B be two groups. We next define their *direct sum* $A \oplus B$. This consists of the set $A \times B$ with addition given by

$$(a_1, b_1) + (a_2, b_2) = (a_1 + a_2, b_1 + b_2).$$

Let us check that $A \oplus B$ satisfies axioms G1 through G4:

G1 $(0, 0) + (a, b) = (0 + a, 0 + b) = (a, b)$.

G2 $(-a, -b) + (a, b) = ((-a) + a, (-b) + b) = (0, 0)$.

G3 $((a_1, b_1) + (a_2, b_2)) + (a_3, b_3) = (a_1 + a_2, b_1 + b_2) + (a_3, b_3)$

$$
\begin{aligned}
&= ((a_1 + a_2) + a_3, (b_1 + b_2) + b_3) \\
&= (a_1 + (a_2 + a_3), b_1 + (b_2 + b_3)) \\
&= (a_1, b_1) + (a_2 + a_3, b_2 + b_3) \\
&= (a_1, b_1) + ((a_2, b_2) + (a_3, b_3))
\end{aligned}
$$

G4 $(a_1, b_1) + (a_2, b_2) = (a_1 + a_2, b_1 + b_2)$

$$= (a_2 + a_1, b_2 + b_1) = (a_2, b_2) + (a_1, b_1).$$

$A \oplus B$ is called the external direct sum, but we loosely refer to any isomorphic group as a direct sum (or product) of A and B. By induction—or analogy—one can define the direct sum of any finite set of (abelian) groups.

We next describe homomorphisms involving direct sums. The result should be compared with C4 of Chapter 1.

2.2 Lemma *Let A, B, X be (abelian) groups; let the map $\phi: X \to A \oplus B$ have components $\phi_1 : X \to A$, $\phi_2 : X \to B$, that is,*

$$\phi(x) = (\phi_1(x), \phi_2(x)) \qquad \text{for all} \quad x \in X.$$

Then ϕ is a homomorphism if and only if ϕ_1 and ϕ_2 both are.

For

$$\phi(x + y) = (\phi_1(x + y), \phi_2(x + y)),$$

and

$$
\begin{aligned}
\phi(x) + \phi(y) &= (\phi_1(x), \phi_2(x)) + (\phi_1(y), \phi_2(y)) \\
&= (\phi_1(x) + \phi_1(y), \phi_2(x) + \phi_2(y)).
\end{aligned}
$$

By definition, ϕ is a homomorphism if and only if, for all $x, y \in X$, the left-hand sides are equal; equality of the two right-hand sides is equivalent to both ϕ_1 and ϕ_2 being homomorphisms. ■

This lemma can be interpreted as defining a bijection

$$\text{Hom}\,(X, A \oplus B) \cong \text{Hom}\,(X, A) \times \text{Hom}\,(X, B).$$

It makes sense to have a symbol for the map ϕ with components ϕ_1 and ϕ_2. We write it as a column matrix—or alternatively (for typographical ease)—as a row matrix in braces instead of parentheses, thus:

$$\phi = \begin{pmatrix} \phi_1 \\ \phi_2 \end{pmatrix} = \{\phi_1, \phi_2\}.$$

This result has a counterpart for homomorphisms from a direct sum.

2.3 Lemma *Let A, B, Y be abelian groups. For any homomorphisms* $\psi_1 : A \to Y$, $\psi_2 : B \to Y$, *the map* $(\psi_1, \psi_2) : A \oplus B \to Y$ *defined by*

$$(\psi_1, \psi_2)(a, b) = \psi_1(a) + \psi_2(b)$$

is a homomorphism. Conversely, every homomorphism $\psi : A \oplus B \to Y$ *can be (uniquely) expressed as a* (ψ_1, ψ_2).

Caution. Lemma 2.2 is analogous to C4 (for continuous maps): it holds for nonabelian as well as abelian groups, and there is a corresponding result for maps of sets. But Lemma 2.3 holds for abelian groups only.

Proof. We leave to the reader to check that (ψ_1, ψ_2) is a homomorphism: notice that you have to use the fact that Y is abelian. Now given any ψ, define ψ_1 and ψ_2 by

$$\psi_1(a) = \psi(a, 0),$$
$$\psi_2(b) = \psi(0, b).$$

It is easily seen that these are homomorphisms, and we have

$$\begin{aligned} \psi(a, b) &= \psi((a, 0) + (0, b)) \\ &= \psi(a, 0) + \psi(0, b) \\ &= \psi_1(a) + \psi_2(b) = (\psi_1, \psi_2)(a, b) \end{aligned}$$

so $\psi = (\psi_1, \psi_2)$, and clearly this decomposition is unique. ∎

The above notation generalizes the usual matrix notation. For observe that the composite of two maps

$$X \xrightarrow{\{\phi_1, \phi_2\}} A \oplus B \xrightarrow{(\psi_1, \psi_2)} Y$$

is given by

$$(\psi_1, \psi_2)\begin{pmatrix} \phi_1 \\ \phi_2 \end{pmatrix} = (\psi_1\phi_1 + \psi_2\phi_2),$$

for

$$\begin{aligned} (\psi_1, \psi_2)\{\phi_1, \phi_2\}x &= (\psi_1, \psi_2)(\phi_1(x), \phi_2(x)) \\ &= \psi_1\phi_1(x) + \psi_2\phi_2(x) \\ &= (\psi_1\phi_1 + \psi_2\phi_2)(x). \end{aligned}$$

This illustrates the usual matrix multiplication law in a special case. Observe now that a map

$$\phi : X \oplus Y \to A \oplus B$$

is of the form $\{\phi_1, \phi_2\}$, where $\phi_1 : X \oplus Y \to A$ and $\phi_2 : X \oplus Y \to B$. We can continue, and write $\phi_1 = (\alpha_{11}, \alpha_{12})$ and $\phi_2 = (\alpha_{21}, \alpha_{22})$. It then seems obvious that ϕ should be denoted by the matrix

$$\begin{pmatrix} \alpha_{11} & \alpha_{12} \\ \alpha_{21} & \alpha_{22} \end{pmatrix}.$$

This is confirmed by the fact that if

$$\psi_1 = \{\alpha_{11}, \alpha_{21}\} : X \to A \oplus B$$

and

$$\psi_2 = \{\alpha_{12}, \alpha_{22}\} : Y \to A \oplus B,$$

then $\phi = (\psi_1, \psi_2)$. The reader can check that the usual matrix multiplication laws are valid in this sort of case, too.

If you are muddled about which maps go where, you may find it useful to remember that the columns of the matrix correspond to the summands X, Y of the domain and the rows to A and B, thus:

$$\begin{matrix} X & Y & \\ \downarrow & \downarrow & \\ \alpha_{11} & \alpha_{12} & \to A \\ \alpha_{21} & \alpha_{22} & \to B. \end{matrix}$$

The notation indicates that, for example, $\alpha_{12} : Y \to A$.

Examples

1 The direct sum of n groups, each a copy of the additive group $\mathbf{R}$ of real numbers, is the vector space $\mathbf{R}^n$, for an element of either is a sequence $(a_1, \ldots, a_n)$ of n real numbers, and in each case we add two sequences by adding their components.

2 A linear map $\mathbf{R} \to \mathbf{R}$ is multiplication by a number a, and is usually specified by a. The above then leads to the usual matrix notation for linear maps $\mathbf{R}^m \to \mathbf{R}^n$.

More relevant to our purposes below, however, is the subgroup $\mathbf{Z}^n$ of $\mathbf{R}^n$: the direct sum of n groups, each a copy of $\mathbf{Z}$. Matrix notation is applicable here too, particularly since by the corollary to Proposition 2.9,

$$\operatorname{Hom}(\mathbf{Z}, \mathbf{Z}) \cong \mathbf{Z}.$$

The next result will be important later on when we want to use a group to define a number.

2.4 Proposition *If $m \neq n$ are positive integers, the groups $\mathbf{Z}^m$ and $\mathbf{Z}^n$ are not isomorphic.*

Proof. First let A be any (additive) abelian group. Define a relation on the elements of A by:

$$a\,R\,b \quad \text{if there exists} \quad c \in A \quad \text{with} \quad a - b = c + c.$$

This is an equivalence relation. For

$a\,R\,a$ since $a - a = 0 = 0 + 0$;

$a\,R\,b$ implies that we can write $a - b = c + c$; but then

$$b - a = -(a - b) = (-c) + (-c),$$

so $b\,R\,a$;

$a\,R\,a'$ and $a'\,R\,a''$ imply that we can write

$$a - a' = b + b \qquad \text{and} \qquad a' - a'' = b' + b'.$$

But then

$$a - a'' = (a - a') + (a' - a'') = (b + b) + (b' + b') = (b + b') + (b + b')$$

so that $a\,R\,a''$.

Thus A is divided up into equivalence classes. Denote by $t(A)$ the number of such classes. If $A = \mathbf{Z}^n$, we find at once that $(r_1, \ldots, r_n)\,R\,(s_1, \ldots, s_n)$ if and only if each $r_i - s_i$ is even. Since there are two possibilities (even or odd) for each r_i, it follows that $t(\mathbf{Z}^n) = 2^n$.

Now if $\phi\colon A \to B$ is a homomorphism and $a, a' \in A$, the definition shows that if $a\,R\,a'$, then $\phi(a)\,R\,\phi(a')$. Thus if ϕ is an isomorphism, $a\,R\,a'$ if and only if $\phi(a)\,R\,\phi(a')$. In this case, ϕ maps each equivalence class in A to one in B, and vice versa. So $t(A) = t(B)$.

Finally if $m \neq n$, then $t(\mathbf{Z}^m) = 2^m \neq 2^n = t(\mathbf{Z}^n)$, so there can be no isomorphism of $\mathbf{Z}^m$ on $\mathbf{Z}^n$. ■

EXACT SEQUENCES

Given three groups and two homomorphisms forming a sequence

$$A_1 \xrightarrow{\phi_1} A_2 \xrightarrow{\phi_2} A_3,$$

the sequence is said to be *exact* if $\operatorname{Ker} \phi_2 = \operatorname{Im} \phi_1$. This can be regarded as comprising two assertions:

i) $\operatorname{Im} \phi_1 \subset \operatorname{Ker} \phi_2$,
which means that for any y of the form $\phi_1(x)$, we have $\phi_2(y) = 0$; or equivalently, for any $x \in A_1$, that $\phi_2\phi_1(x) = 0$; or more briefly, that $\phi_2\phi_1 = 0$.

ii) $\operatorname{Ker} \phi_2 \subset \operatorname{Im} \phi_1$,
which means that if $y \in A_2$ satisfies $\phi_2(y) = 0$, then we can find $x \in A_1$ with $y = \phi_1(x)$.

A sequence of groups and homomorphisms is called exact if any two consecutive homomorphisms of the sequence satisfy the condition above. In such a case the assertion above will be expressed by saying that the longer sequence is *exact at* A_2.

Examples Let B be a subgroup of A. Then $\{0\} \to B \subset A$ is exact. More generally, by Lemma 2.1, $\{0\} \to A \xrightarrow{\phi} X$ is exact if and only if ϕ is injective, and $A \xrightarrow{\phi} X \to \{0\}$ is exact if and only if ϕ is surjective. Thus $\{0\} \to A \xrightarrow{\phi} X \to \{0\}$ is exact only if ϕ is bijective. For any abelian groups A, B, the sequence

$$\{0\} \to A \xrightarrow{\{1,0\}} A \oplus B \xrightarrow{(0,1)} B \to \{0\}$$

is exact.

The next result goes in the opposite direction.

2.5 Theorem *Given an exact sequence*

$$P \xrightarrow{\alpha} Q \xrightarrow{\beta} R \xrightarrow{\gamma} S \to \{0\}$$

and a homomorphism $\delta: S \to R$ *with* $\gamma\delta = 1$, *then the sequence*

$$P \xrightarrow{\{\alpha,0\}} Q \oplus S \xrightarrow{(\beta,\delta)} R \to \{0\}$$

is exact.

Proof. There are three things to check, since we must prove the sequence exact at $Q \oplus S$ and at R.

$\operatorname{Im}\{\alpha, 0\} \subset \operatorname{Ker}(\beta, \delta)$: We have $(\beta, \delta)\{\alpha, 0\} = \beta\alpha = 0$.

$\operatorname{Ker}(\beta, \delta) \subset \operatorname{Im}\{\alpha, 0\}$: Let $(\beta, \delta)(q, s) = 0$. Then

$$0 = \gamma(\beta, \delta)(q, s) = (\gamma\beta, \gamma\delta)(q, s) = (0, 1)(q, s) = s,$$

and so also, substituting $s = 0$,

$$0 = (\beta, \delta)(q, 0) = \beta(q).$$

By exactness at Q of the given sequence, there exists p with $q = \alpha(p)$. But then

$$(q, s) = (q, 0) = (\alpha(p), 0) = \{\alpha, 0\}(p).$$

$\operatorname{Im}(\beta, \delta) = R$: Since $\gamma\delta = 1$, we have, for any $r \in R$, $\gamma(r) = \gamma\delta\gamma(r)$. Thus $r - \delta\gamma(r) \in \operatorname{Ker}\gamma$. But by exactness (at R) of the given sequence, $\operatorname{Ker}\gamma = \operatorname{Im}\beta$. Thus for some $q \in Q$,

$$r - \delta\gamma(r) = \beta(q).$$

And now

$$r = (\beta, \delta)(q, \gamma(r)). \blacksquare$$

Corollary *Given an exact sequence*

$$\{0\} \to Q \xrightarrow{\beta} R \xrightarrow{\gamma} S \to \{0\}$$

and $\delta: S \to R$ *with* $\gamma\delta = 1$, *the map* $(\beta, \delta): Q \oplus S \to R$ *is an isomorphism.*

For putting $P = 0$ in Theorem 2.5, we get an exact sequence

$$\{0\} \to Q \oplus S \xrightarrow{(\beta, \delta)} R \to \{0\};$$

the result now follows by a remark above. $\blacksquare$

When a homomorphism exists such as δ in the above corollary, the exact sequence is said to *split*, and we also say that R splits as a direct sum of Q and S.

The next result has less theoretical value, but will be very useful nonetheless.

2.6 Theorem *Given an exact sequence*

$$X \xrightarrow{\{\alpha, \beta\}} A \oplus B \xrightarrow{\begin{pmatrix} a & b \\ c & d \end{pmatrix}} A' \oplus B' \xrightarrow{(\alpha', \beta')} X',$$

such that $d: B \to B'$ *is an isomorphism, then the sequence*

$$X \xrightarrow{\alpha} A \xrightarrow{a - bd^{-1}c} A' \xrightarrow{\alpha'} X'$$

is exact, and also $\operatorname{Ker}\alpha = \operatorname{Ker}\{\alpha, \beta\}$, $\operatorname{Im}\alpha' = \operatorname{Im}(\alpha', \beta')$.

Proof

$$0 = \begin{pmatrix} a & b \\ c & d \end{pmatrix}\begin{pmatrix} \alpha \\ \beta \end{pmatrix} = \begin{pmatrix} a\alpha + b\beta \\ c\alpha + d\beta \end{pmatrix}.$$

Since d is an isomorphism, it has an inverse. So from $0 = c\alpha + d\beta$ we deduce $0 = d^{-1}(c\alpha + d\beta) = d^{-1}c\alpha + \beta$, that is $\beta = -d^{-1}c\alpha$. It follows that for $x \in X$, $\alpha x = 0$ implies $\beta x = 0$, so $\alpha x = 0$ if and only if αx and βx are both zero; that is $\operatorname{Ker} \alpha = \operatorname{Ker} \{\alpha, \beta\}$.

We also have $a\alpha + b\beta = 0$ or, substituting for β,

$$0 = a\alpha + b(-d^{-1}c\alpha) = (a - bd^{-1}c)\alpha,$$

proving one part of exactness at A. For the other part, suppose $u \in A$ such that $(a - bd^{-1}c)u = 0$. Write $v = -d^{-1}c(u)$, so that $c(u) + d(v) = 0$, and also $a(u) + b(v) = 0$. Thus (u, v) is in the kernel of the middle map in the given sequence; by exactness it is also in the image of $\{\alpha, \beta\}$. So for some $x \in S$ we have $\alpha(x) = u$ (also $\beta(x) = v$), proving exactness at A.

The proof of the other half of Theorem 2.6 is very similar to this. First,

$$0 = (\alpha', \beta')\begin{pmatrix} a & b \\ c & d \end{pmatrix} = (\alpha'a + \beta'c, \alpha'b + \beta'd),$$

so $\beta' = -\alpha'bd^{-1}$ and $\alpha'(a - bd^{-1}c) = 0$. The second statement gives half of exactness at A'; the first implies that $\operatorname{Im} \alpha' = \operatorname{Im} (\alpha', \beta')$. For clearly $\operatorname{Im} \alpha' \subset \operatorname{Im} (\alpha', \beta')$; conversely any element of $\operatorname{Im} (\alpha', \beta')$ is of the form

$$\alpha'(p) + \beta'(q) = \alpha'\big(p - bd^{-1}(q)\big),$$

so belongs to $\operatorname{Im} \alpha'$. Finally, if $p \in A'$ with $\alpha'(p) = 0$, then

$$(\alpha', \beta')(p, 0) = \alpha'(p) = 0,$$

so by exactness of the given sequence there exists $(u, v) \in A \oplus B$ with

$$\begin{pmatrix} a & b \\ c & d \end{pmatrix}(u, v) = (p, 0),$$

that is,

$$a(u) + b(v) = p,$$

$$c(u) + d(v) = 0,$$

so $v = -d^{-1}c(u)$ and $p = (a - bd^{-1}c)(u)$. ■

Corollary *If*

$$\begin{pmatrix} a & b \\ c & d \end{pmatrix} : A \oplus B \to A' \oplus B'$$

is an isomorphism, and $d: B \to B'$ is an isomorphism, then A and A' are isomorphic; in particular if b (or c) is 0, a is an isomorphism.

This follows at once from Theorem 2.6 if we take $X = X' = \{0\}$. ■

FREE ABELIAN GROUPS

We are now ready for the final topic in this chapter: the free abelian group. We will introduce this axiomatically, using a crucial property.

Definition *Given a set X, an abelian group G, and a map $i: X \to G$. Then (G, i) has the universal property for X if for any abelian group A and map $j: X \to A$, there exists a unique homomorphism $\phi: G \to A$ with $j = \phi \circ i$.*

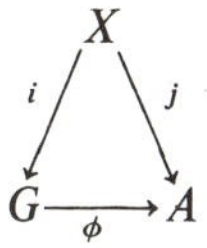

When this is so, we also say that G is the free abelian group on generators $i(X)$. The cardinal number of X is called the rank of G.

Before discussing examples, we give one application of this property to a question arising above.

2.7 Proposition *Any short exact sequence*

$$0 \to A \xrightarrow{i} G \xrightarrow{p} B \to 0,$$

with B free abelian, splits.

Proof. Let $b: X \to B$ be such that (B, b) has the universal property for X. For each $x \in X$, choose $g(x) \in G$ with $p(g(x)) = b(x)$; this is possible since p is surjective.

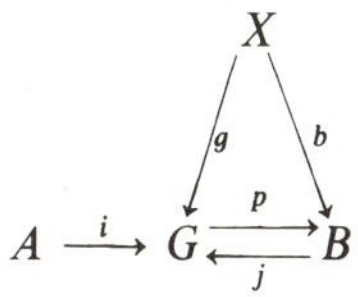

By the universal property, there is a homomorphism $j: B \to G$ with $j \circ b = g$. Then

$$p \circ j \circ b = p \circ g = b.$$

By the uniqueness part of the universal property, it follows that the homomorphism $p \circ j$ coincides with 1_B. Thus the sequence is split. ■

As another illustration of how to use the property, we prove the essential uniqueness of (G, i) having the universal property for a given X. We will then consider existence.

2.8 Proposition *Suppose (G, i) and (H, j) both have the universal property for X. Then there is a unique isomorphism $\phi: G \to H$ with $j = \phi \circ i$.*

Proof. By the universal property we obtain such a homomorphism ϕ; similarly $\psi: H \to G$ with $i = \psi \circ j$. Now consider the diagram

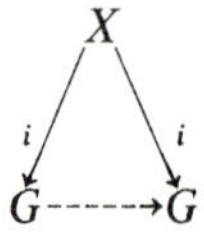

By the universal property, there is a unique homomorphism $\chi: G \to G$ with $i = \chi \circ i$. But we already have two examples of such χ: 1_G (clearly) and $\psi \circ \phi$, since $\psi \circ \phi \circ i = \psi \circ j = i$. Hence $\psi \circ \phi = 1_G$. Similarly $\phi \circ \psi = 1_H$. Hence ϕ is injective and surjective, and so bijective. ■

We must now consider the problem of existence of free abelian groups. Let us begin with the easiest case.

2.9 Proposition *Let X have only one element x. Define $i: X \to \mathbf{Z}$ by $i(x) = 1$. Then $(\mathbf{Z}, i)$ has the universal property for X.*

Proof. Let A be an (abelian) group, $j: X \to A$; write $j(x) = a$. We have to prove that there is a unique homomorphism $\phi: \mathbf{Z} \to A$ with $j = \phi \circ i$, that is, with $a = j(x)$ equal to $\phi(i(x)) = \phi(1)$.

Now since ϕ is a homomorphism,

$$\phi(2) = \phi(1 + 1) = \phi(1) + \phi(1) = a + a,$$

$$\phi(3) = \phi(2 + 1) = \phi(2) + \phi(1) = (a + a) + a.$$

This suggests that for n a positive integer, we define $\phi(n)$ by induction, as

F1 $\phi(0) = 0,$

F2 $\phi(n+1) = \phi(n) + a$,

and then for the set of negative numbers,

F3 $\phi(-n) = -\phi(n)$.

Certainly no other map ϕ can have the desired properties; it remains to check that ϕ as so defined is a homomorphism.

Now F2 holds by definition for $n \geqslant 0$. Let us check that it is true for all $n \in \mathbf{Z}$. For $n \geqslant 0$ we have

$$\begin{aligned} \phi(n+1) &= \phi(n) + a \\ \phi(-n-1) &= -\phi(n+1) = -(\phi(n) + a) \\ &= -\phi(n) - a \\ &= \phi(-n) - a, \end{aligned}$$

and so $$\phi(-n) = \phi(-n-1) + a,$$

that is F2 is true also for $n < 0$. We can rewrite this as

$$\phi(n+1) = \phi(n) + \phi(1).$$

Next, by induction, check

$$\phi(n+m) = \phi(n) + \phi(m) \qquad \text{for} \quad m \geqslant 0.$$

This is clear for $m = 0$; if it holds for $m = k$, then

$$\begin{aligned} \phi(n+k+1) &= \phi(n+k) + \phi(1) \\ &= \phi(n) + \phi(k) + \phi(1) \\ &= \phi(n) + \phi(k+1). \end{aligned}$$

So it holds for $m = k + 1$. But it holds for $m = -k$ also, since

$$\begin{aligned} \phi(n-k) &= -\phi(-n+k) \\ &= -(\phi(-n) + \phi(k)) \\ &= -\phi(-n) - \phi(k) \\ &= \phi(n) + \phi(-k). \end{aligned}$$

This completes the proof that ϕ is a homomorphism, and hence proves Proposition 2.9. ■

In future we shall write $n \cdot a$ for the element $\phi(n) \in A$ defined above.

Corollary *For any abelian group A, the map*

$$\mathrm{ev}: \mathrm{Hom}(\mathbf{Z}, A) \to A$$

defined for $f: \mathbf{Z} \to A$ *by* $\mathrm{ev}(f) = f(1)$ *is an isomorphism.*

For by the definition H2 of addition of homomorphisms, ev is a homomorphism of groups. Also ev is bijective since, by the universal property just proved, for any $a \in A$, there is a unique homomorphism $f: \mathbf{Z} \to A$ with $f(1) = a$.

2.10 Proposition *Suppose* (G_1, i_1) *has the universal property for* X_1, *and* (G_2, i_2) *for* X_2. *Let* X_1, X_2 *be disjoint. Define*

$$i: X_1 \cup X_2 \to G_1 \oplus G_2$$

by

$$i(x) = (i_1(x), 0) \quad \text{if} \quad x \in X_1,$$
$$i(x) = (0, i_2(x)) \quad \text{if} \quad x \in X_2.$$

Then $(G_1 \oplus G_2, i)$ *has the universal property for* $X_1 \cup X_2$.

Proof. Given an abelian group A and map $j: X \to A$; since (G_1, i_1) has the universal property for X_1, there is a unique homomorphism $\phi_1 : G_1 \to A$ with $j|X_1 = \phi_1 \circ i_1$. Similarly, we have $\phi_2: G_2 \to A$ with $j|X_2 = \phi_2 \circ i_2$. Then if

$$\phi = (\phi_1, \phi_2): G \oplus G_2 \to A,$$

we have

$$\phi i(x) = \phi(i_1(x), 0) = \phi_1 i_1(x) = j(x) \qquad \text{for} \quad x \in X_1;$$

similarly for $x \in X_2$, so $\phi \circ i = j$. Conversely, if $\psi = (\psi_1, \psi_2)$ satisfies $\psi \circ i = j$, we deduce by the same argument that $\psi_1 \circ i_1 = j|X_1$, so by the universal property of i_1, $\psi_1 = \phi_1$. Similarly $\psi_2 = \phi_2$. ■

The same argument now works by induction for finite direct sums. It thus follows, using Proposition 2.9, that if X is finite, there exists (G, i) with the universal property for X, and Proposition 2.8 already gives us uniqueness. We now describe the group G given by the construction. Let X have n elements, $x_1, x_2, \ldots, x_n$. We must form the direct sum of n groups, each a copy of the group $\mathbf{Z}$ of integers. An element of this direct sum is a sequence of n integers $(r_1, r_2, \ldots, r_n)$, which can be thought of as a map from X to $\mathbf{Z}$ (sending x_i to r_i)—or we can think of the elements as the points of $\mathbf{R}^n$ with integer coordinates. The group structure can then be thought of as vector addition in $\mathbf{R}^n$:

$$(r_1, \ldots, r_n) + (s_1, \ldots, s_n) = (r_1 + s_1, \ldots, r_n + s_n).$$

We write $\mathbf{Z}^n$ for this group. Then $(\mathbf{Z}^n, i)$ has the universal property for X, where $i: X \to \mathbf{Z}^n$ has $i(x_k)$ the unit point on the kth coordinate axis, or alternatively, using Kronecker's δ-notation,

$$i(x_k) = (\delta_{k1}, \delta_{k2}, \ldots, \delta_{kn}).$$

Thus $\mathbf{Z}^n$ is the free abelian group of (finite) rank n. By Proposition 2.4, the rank is determined once we know the abstract group structure. We now wish to construct free abelian groups of arbitrary (infinite) rank. We use the above description of $\mathbf{Z}^n$ as a model.

Let X be any set. Write $F(X)$ for the set of those maps $\xi: X \to \mathbf{Z}$ with $\xi(X) = 0$ for all but a finite number of elements x of X. Given $\xi, \eta \in F(x)$ we define $-\xi, (\xi + \eta)$ by

$$\left.\begin{aligned} (-\xi)(x) &= -\xi(x) \\ (\xi + \eta)(x) &= \xi(x) + \eta(x) \end{aligned}\right\} \quad \text{for all} \quad x \in X.$$

It is easy to verify that with these operations, $F(X)$ is an abelian group. Define $i_X: X \to F(X)$ by

$$i_X(x)(y) = \begin{cases} 0 & \text{if} \quad y \neq x, \\ 1 & \text{if} \quad y = x. \end{cases}$$

2.11 Theorem $(F(X), i_X)$ *has the universal property for* X.

Proof. First we observe that if $Y \subset X$, we can identify $F(Y)$ with the subgroup of $F(X)$ consisting of these maps $\xi: X \to \mathbf{Z}$ with $\xi(X - Y) = 0$. Now if $\xi \in F(X)$, there is some finite subset Y of X with $\xi(X - Y) = 0$, and so $\xi \in F(Y)$. Thus

$$F(X) = \bigcup \{F(Y): Y \subset X, Y \text{ finite}\}.$$

Next note that the argument following Proposition 2.10 shows, in our present notation, that if Y is finite, then $(F(Y), i_Y)$ has the universal property for Y.

Now let A be any abelian group, and $j: X \to A$ a map. For any finite subset Y of X we know that there is a unique homomorphism $\phi_Y: F(Y) \to A$ with $\phi_Y \circ i_Y = j|Y$. Now, if $Z \subset Y$ and so $F(Z) \subset F(Y)$, then $i_Z = i_Y|Z$, and so $(\phi_Y|F(Z)) \circ i_Z = j|Z$. By uniqueness, $\phi_Z = \phi_Y|F(Z)$. We now define $\phi: F(X) \to A$. For any $\xi \in F(X)$ choose a finite subset Y of X with $\xi \in F(Y)$, and define $\phi(\xi) = \phi_Y(\xi)$. This is independent of Y, for if also $\xi \in F(Y')$, we have

$$\phi_Y(\xi) = \phi_{Y \cup Y'}(\xi) = \phi_{Y'}(\xi)$$

by the above. Also, ϕ is a homomorphism. For, given $\xi_1, \xi_2 \in F(X)$, choose finite subsets Y_1, Y_2 of X with $\xi_1 \in F(Y_1)$, $\xi_2 \in F(Y_2)$, and write $Y = Y_1 \cup Y_2$. Then

$$\phi(\xi_1 + \xi_2) = \phi_Y(\xi_1 + \xi_2) = \phi_Y(\xi_1) + \phi_Y(\xi_2) = \phi(\xi_1) + \phi(\xi_2),$$

since ϕ_Y is a homomorphism. For any $x \in X$, we have

$$\phi\big(i(x)\big) = \phi_{\{x\}}\big(i_{\{x\}}(x)\big) = j(x);$$

thus $\phi \circ i = j$.

Finally, given any homomorphism $\psi: F(X) \to A$ with $\psi \circ i = j$, we have for any finite $Y \subset X$,

$$\psi|F(Y) \circ i_Y = \psi \circ (i|Y) = j|Y = \phi_Y \circ i_Y = \phi|F(Y) \circ i_Y,$$

and hence $\psi|F(Y) = \phi|F(Y)$ by the uniqueness already proved for Y. But since $F(X)$ is the union of these subgroups $F(Y)$, it follows that $\psi = \phi$. ■

For each set X we have constructed $\big(F(X), i_X\big)$ having the universal property for X. We usually write just i for i_X, and call $F(X)$ *the* free abelian group *on* X. Now given any map $f: X \to Y$ of sets, there is (by the universal property) a unique homomorphism $F(f): F(X) \to F(Y)$ such that $F(f) \circ i_X = i_Y \circ f$, or, in diagrammatic form,

$$\begin{array}{ccc} X & \xrightarrow{f} & Y \\ \downarrow{\scriptstyle i_X} & & \downarrow{\scriptstyle i_Y} \\ F(X) & \xrightarrow{F(f)} & F(Y). \end{array}$$

FURTHER DEVELOPMENTS

As in Chapter 1, we have barely scratched the surface of a large and flourishing branch of pure mathematics. It would be invidious for a topologist to select for recommendation from the plethora of introductory books on group theory. Rather, let me mention that our viewpoint here (as later in the book) derives from an even more abstract branch of mathematics, category theory. Introductions (not too elementary) are given in the stimulating books by Freyd and Mitchell.

Freyd, P. *Abelian Categories*, Harper and Row, New York, 1964.

Mitchell, B. *Theory of Categories*, Academic Press, New York, 1965.

EXERCISES AND PROBLEMS

1. Verify that **Z**, **R**, **C** are additive and that $\mathbf{R}^*$, $\mathbf{C}^*$, and S^1 are multiplicative groups.
2. Show that if ϕ is a homomorphism of additive groups, then $\phi(0) = 0$ and $\phi(-a) = -\phi(a)$.
3. a) Prove Lemma 2.1.
   b) Justify the assertions in the example following Lemma 2.1.
4. A, B are subgroups of G. Show that $A \cap B$ is a subgroup, and give an example to show that $A \cup B$ is not. In fact, show that the smallest subgroup containing A and B is
$$A + B = \{a + b : a \in A, b \in B\}.$$
5. Suppose, given an exact sequence,
$$0 \to A \xrightarrow{i} G \xrightarrow{p} B \to 0.$$
   a) Given a splitting homomorphism $j: B \to G$ with $p \circ j = 1_B$, show that there is a unique homomorphism $q: G \to A$ with $q \circ i = 1_A$ and $q \circ j = 0$.
   b) Given a homomorphism $q: G \to A$ with $q \circ i = 1_A$, show that there is a unique $j: B \to G$ with $p \circ j = 1_B$ and $q \circ j = 0$.
   c) Show that when (a) and (b) hold, $j \circ p + i \circ q = 1_G$.
6. Let $i: A \to G$ and $q: G \to A$ satisfy $q \circ i = 1_A$. Show that G is the direct sum of $i(A)$ and Ker q.
7. Given groups and homomorphisms
$$i: A \to G, \qquad j: B \to G, \qquad q: G \to A, \qquad p: G \to B$$
   satisfying $p \circ i = 0$, $q \circ j = 0$, $p \circ j = 1_B$, $q \circ i = 1_A$ and $j \circ p + i \circ q = 1_G$, show that the sequences
$$0 \to A \xrightarrow{i} G \xrightarrow{p} B \to 0,$$
$$0 \to B \xrightarrow{j} G \xrightarrow{q} A \to 0$$
   are exact.
8. With the notation of the preceding exercise, show (by constructing inverse maps or otherwise) that for any group X,
$$f \mapsto (g \circ f, p \circ f)$$
   induces a bijection
$$\text{Hom}\,(X, G) \to \text{Hom}\,(X, A) \times \text{Hom}\,(X, B)$$
   and that $f \mapsto (i \circ f, j \circ f)$ induces a bijection
$$\text{Hom}\,(G, X) \to \text{Hom}\,(A, X) \times \text{Hom}\,(B, X).$$
   Show also that these bijections are group isomorphisms.
9. a) Given a homomorphism $j: G_1 \to H$ and a surjective homomorphism $\alpha: G_1 \to G_2$,

show that there exists a homomorphism $\beta: G_2 \to H$ with $\beta \circ \alpha = \gamma$ if and only if $\alpha(\text{Ker } \gamma) = 0$.

b) Give an example to show that if α is not surjective, the condition $\alpha(\text{Ker } \gamma) = 0$ does not imply the existence of β.

*10. Give an example of an exact sequence which is not split.

11. Given a set X, and $x \in X$, and $\varepsilon: X \to \mathbf{Z}$ such that $\varepsilon(x) = 1$. Let $p: F(X) \to \mathbf{Z}$ satisfy $p \circ i_X = \varepsilon$. Show that p is a split surjection. Show also that if ε_1, p_1 and ε_2, p_2 are as above, there is a unique isomorphism $\phi: \text{Ker } p_1 \to \text{Ker } p_2$ such that for all $\xi \in \text{Ker } (p_1)$, $\phi(\xi) - \xi$ is a multiple of $i_X(x)$.

12. Give an alternative proof of Theorem 2.6, by representing any homomorphism $\phi: \mathbf{Z}^m \to \mathbf{Z}^n$ by a matrix of coefficients, and showing that if ϕ is an isomorphism, the matrix is invertible.

13. Let (G, i) have the universal property for X. Show that for any group A there is a bijection of Map (X, A) and Hom (G, A).

14. Let (G, i) satisfy the universal property for X, *except* for the uniqueness. Construct a homomorphism $F(X) \to G$, and show that G splits as a direct sum $F(X) \oplus A$. Prove also the converse. [*Hint.*· Use Exercise 5.]

15. Show that
$$\text{Ker } \{\alpha, \beta\} = \text{Ker } \alpha \cap \text{Ker } \beta,$$
$$\text{Im } (\alpha, \beta) = \text{Im } \alpha + \text{Im } \beta.$$

16. Suppose that for each i, C_i is a free abelian group of rank α_i.

a) If $0 \to C_0 \to C_1 \to C_2 \to 0$ is exact, show that $\alpha_0 + \alpha_2 = \alpha_1$. [*Hint.*· Use Theorem 2.5.]

b) Given that
$$0 \to C_n \to C_{n-1} \to \cdots \to C_1 \to C_0 \to 0$$
is exact, show that
$$\sum_0^n (-1)^i \alpha_i = 0.$$

PART 1
INTRODUCTION TO HOMOTOPY THEORY

3
CONNECTED AND DISCONNECTED SPACES

INTRODUCTION

We begin our introduction to topology with the study of connectedness—traditionally the only topic studied in both analytic and algebraic topology. Our approach here is descriptive; later, we will be more formal, and will also generalize the notion. Note that the Jordan curve theorem is a statement about connectedness properties of the complement of a Jordan curve in the plane.

CONNECTEDNESS

Let X be a space. Suppose X expressed as the disjoint union of nonempty subsets U and V:

$$U \cup V = X, \qquad U \cap V = \phi.$$

Suppose, in addition, that U and V are both open subsets of X. Then (U, V) is called a *partition* of X, and X is *disconnected.* If X has no partition, it is *connected.*

There are several equivalent forms of this definition. First of all, note that U is the complement of V in X, so the condition that U be open is equivalent to the condition that V be closed, and vice versa. Next, we note that the simplest example of a disconnected space is a space with just two points.

3.1 Lemma *The subset* $\{0, 1\}$ *of* **R** *is disconnected.*

Proof. Take $U = \{0\}$, $V = \{1\}$. Clearly these are disjoint and nonempty, and their union is our space. Also each is open: U is the intersection of our space with the open interval $]-\frac{1}{2}, \frac{1}{2}[$, and similarly for V. ■

Not only is $\{0, 1\}$ disconnected, but it typifies all disconnected spaces.

For let (U, V) be a partition of X. We define a function on X with values 0 and 1 by

$$f(U) = 0,$$

$$f(V) = 1.$$

Then f is continuous: in fact for any subset S of $\{0, 1\}$, $f^{-1}(S)$ is open. For there are only four subsets S: ϕ, $\{0\}$, $\{1\}$, and $\{0, 1\}$ itself. The inverse images of these are, respectively, ϕ, U, V, and X; and all of these are open by hypothesis. Conversely, let $f: X \to \{0, 1\}$ be a continuous map which takes both the values 0 and 1. We write $U = f^{-1}\{0\}$, $V = f^{-1}\{1\}$. Then U and V are open in X since f is continuous and $\{0\}$ and $\{1\}$ are open in $\{0, 1\}$ (see proof of Lemma 3.1). Each is nonempty by hypothesis and, clearly,

$$U \cup V = X, \qquad U \cap V = \phi.$$

This completes the proof of

3.2 Lemma *This space X is disconnected if and only if there exists a continuous surjective map of X onto the space $\{0, 1\}$.* ■

It is quite easy to show that certain spaces are disconnected.

Example *The hyperbola H defined by the equation $x^2 - y^2 = 1$ is disconnected.*

Proof. The hyperbola contains the points $(1, 0)$ and $(-1, 0)$. However, if we substitute $x = 0$, we find $y^2 = -1$ which has no (real) solution. We take

$$U = \{(x, y) \in H : x > 0\},$$

$$V = \{(x, y) \in H : x < 0\}.$$

The first clause of the proof shows that each is nonempty, and the second that $U \cup V = X$; it is clear that $U \cap V = \phi$. Also, U and V are defined as the intersections of X with sets (defined by $x > 0$ and $x < 0$, respectively) which are open in the plane. So U and V are open in X. Hence we have a partition $X = (U, V)$, and X is disconnected.

It is not nearly so easy to show directly that spaces are connected; instead, we need a more roundabout argument. The most important example is an interval.

3.3 Theorem *The interval $[0, 1]$ of real numbers is connected.*

Proof. We suppose not, and will obtain a contradiction. Then there is a continuous function, f say, on $[0, 1]$ taking only the two values 0 and 1. We suppose $f(1) = 1$ [if not, replace $f(x)$ by $g(x) = 1 - f(x)$]. By hypothesis, f

takes the value 0. Let ξ be the least upper bound of the points x with $f(x) = 0$. The contradiction will be obtained by looking at $f(\xi)$.

Since f is continuous at ξ, there exists a number $\delta > 0$ such that if $0 \leqslant x \leqslant 1, |x - \xi| < \delta$, we have

$$|f(x) - f(\xi)| < 1.$$

As f takes only the values 0 and 1, this implies $f(x) = f(\xi)$. Now $f(x) = 1$ for $\xi < x \leqslant 1$, so $f(\xi) = 1$ (if $\xi = 1$, this holds by hypothesis). On the other hand, ξ is the least upper bound of x with $f(x) = 0$, so we can find some such x in the interval $(\xi - \delta, \xi)$. Hence $f(\xi) = 0$ [if $\xi = 0$, this holds since $f(x) = 1$ for $0 < x \leqslant 1$ and f takes both values by hypothesis]. Now we have our contradiction: the theorem is proved. ∎

PATH-CONNECTEDNESS

By looking at the interval we are led to a rather different notion of connectedness. We will call a space X *path-connected* if for any two points $x, y \in X$ we can find a continuous map $f: I \to X$ with $f(0) = x$, $f(1) = y$. In general, a continuous map $f: I \to X$ is called a path in X, and we say that the path above joins x to y.

Examples An ellipse is path-connected. So are the following sets:

$$D^n = \{x \in \mathbf{R}^n : |x| \leqslant 1\},$$

$$S^{n-1} = \{x \in \mathbf{R}^n : |x| = 1\} \qquad (n \geqslant 2).$$

For we can choose coordinates with respect to which the ellipse E has the equation

$$x^2/a^2 + y^2/b^2 = 1.$$

Now define a map $f: \mathbf{R} \to E$ by

$$f(\theta) = (a\cos\theta, b\sin\theta).$$

This map is continuous (the sine and cosine functions are) and, as is well known, surjective. So two points of E can be taken as $f(\theta_0)$, $f(\theta_1)$. The desired path is now

$$g(t) = f\big((1 - t)\theta_0 + t\theta_1\big).$$

The last part of this proof is no more than showing that [0, 1] can equally be replaced by any closed interval in $\mathbf{R}$.

Next, we observe merely that D^n is convex, so two points of it can be joined by a straight line segment.

Given two points of S^{n-1}, we draw a plane through them (say for definiteness, through the origin also). This meets S^{n-1} in a circle containing the two points. The path around the circle is now constructed as above.

We introduced path-connectedness by analogy with connectedness; we now study the relation between these two properties.

3.4 Theorem *Every path-connected space is connected. Not every connected space is path-connected.*

Proof. First suppose that X is path-connected, but not connected. Let $f: X \to \{0, 1\}$ be a surjective continuous map. Choose points $x, y \in X$ with $f(x) = 0$, $f(y) = 1$. Choose a path $g: I \to X$ with $g(0) = x$, $g(1) = y$. Then $f \circ g: I \to \{0, 1\}$ is a continuous map with $f \circ g(0) = 0$ and $f \circ g(1) = 1$, contradicting Theorem 3.3. This proves the first assertion of the theorem.

For the second, we have to construct a connected space which is not path-connected. The space we use, with various minor modifications, is a fruitful source of all kinds of counterexamples.

Let $Y = U \cup V$, where

$$U = \{(x, y) \in \mathbf{R}^2 : x = 0, -1 \leqslant y \leqslant 1\},$$

$$V = \{(x, y) \in \mathbf{R}^2 : 0 < x \leqslant 1, y = \sin \tfrac{1}{x}\}.$$

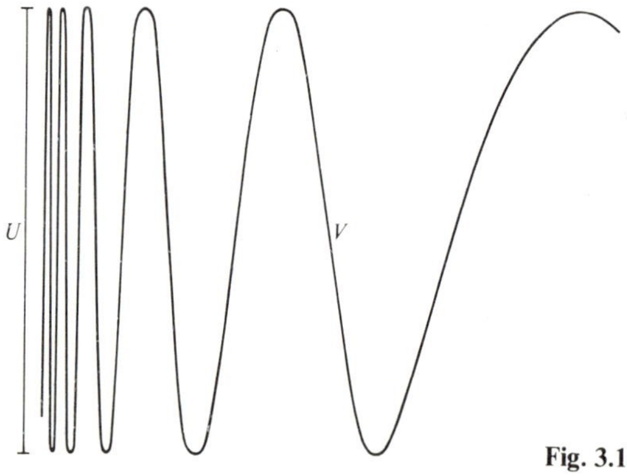

Fig. 3.1

Then U is a closed interval, so is path-connected. V has the parametrization $(x, \sin 1/x)$, and $\sin 1/x$ is a continuous function on $0 < x \leqslant 1$, so V also is path-connected. Thus if Y is disconnected, (U, V) is the only conceivable partition. (Any other would partition the connected space U or V). Define

$f: Y \to \{0, 1\}$ by $f(U) = 0$, $f(V) = 1$. We claim that f is not continuous at any point of U; we will check this at the origin. In fact, $f(0, 0) = 0$ but $f((k\pi)^{-1}, 0) = 1$ for all positive integers k. Any neighborhood of $(0, 0)$ contains some points $((k\pi)^{-1}, 0)$, so f is discontinuous and Y connected.

On the other hand, no path in Y joins a point of U to a point of V. Suppose, in fact, that $f: I \to Y$ has $f(0) \in U$ and $f(1) \in V$. Let ξ be the least upper bound of those $x \in I$ for which $f(x) \in U$. Since f is continuous and U closed (in $\mathbf{R}^2$, hence in Y), $f(\xi) \in U$. Now f is continuous, so for suitable $\delta > 0$, $f[\xi, \xi + \delta]$ lies within unit distance of $f(\xi)$ (Fig. 3.1). Consider the intersection J of V with the disk:

$$\{(x, y) \in \mathbf{R}^2 : d(f(\xi), (x, y)) \leqslant \tfrac{1}{2}\}.$$

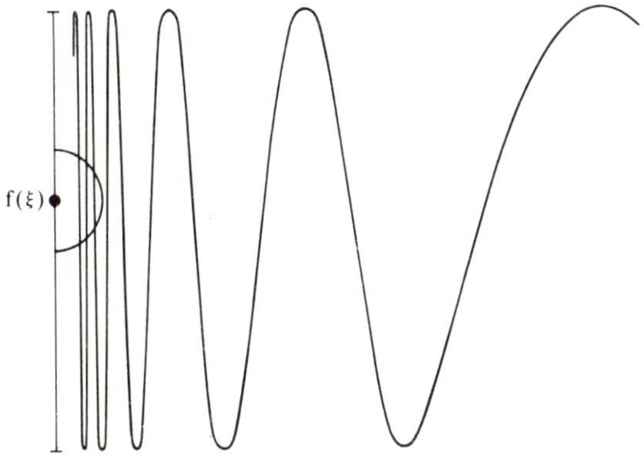

As $\sin 1/x$ oscillates between ± 1 as $x \to 0$, V is "chopped up" in J into an infinite sequence of closed, more-or-less vertical intervals in the disk. If K is the interval containing $f(\xi + \delta)$, then $(K, J - K)$ is a partition of J, so no continuous map of an interval into J can jump from one of these to another. Hence our contradiction, since $f(\xi + \delta) \in K$, $f(\xi) \in J - K$. ■

It may be objected that our counterexample Y is an unreasonably complicated, or "nasty" space, and that "connectedness ought to imply path-connectedness for any reasonable space." This is in fact the case, and the convenient way to express this idea precisely turns out to be as follows.

LOCAL PATH-CONNECTEDNESS

Definition *A topological space X is locally path-connected (l.p.c.) at a if each neighborhood U of a contains a neighborhood V such that any two points of V are connected by a path in U.*

If W is the set of points which can be connected to a by a path in U, then evidently $V \subset W \subset U$ and W is path-connected. So we might equivalently require that every neighborhood of a contain a path-connected neighborhood.

We call X "l.p.c." if it is so at each point. We can construct path-connected spaces by using

3.5 Lemma

i) *$\mathbf{R}^n$ is locally path-connected.*

ii) *If X is l.p.c., U open in X, then U is l.p.c.*

iii) *If the sets U_α are l.p.c. and open in X, and $X = \bigcup_\alpha U_\alpha$, then X is l.p.c.*

Proof

i) Each point of $\mathbf{R}^n$ has arbitrarily small convex neighborhoods (e.g. disks), which are path-connected.

ii) If $a \in U$, and V is a neighborhood of a in U, then (since U is open in X), V is also a neighborhood of a in X. Hence it contains a path-connected neighborhood.

iii) If $x \in X$, then $x \in U_\alpha$ for some α, and any neighborhood N of x in X contains $N \cap U_\alpha$, which contains a path-connected neighborhood M of x in U_α. As U_α is open, M is a neighborhood of x in X. ∎

Parts (ii) and (iii) of Lemma 3.5 formalize the vague idea that local path-connectedness depends only on the behaviour of X near a point, i.e., that it is a local property. Examples of sets which are *not* l.p.c. are:

Examples

1. The set of points

$$P = \{0, \tfrac{1}{n} : n \text{ a positive integer}\} \subset \mathbf{R}$$

is not l.p.c. at 0.

2. The union of the segments in $\mathbf{R}^2$ joining $P \times 0$ to $(0, 1)$ is not l.p.c. at any $(0, y)$ $0 \leqslant y < 1$. See Fig. 3.2.

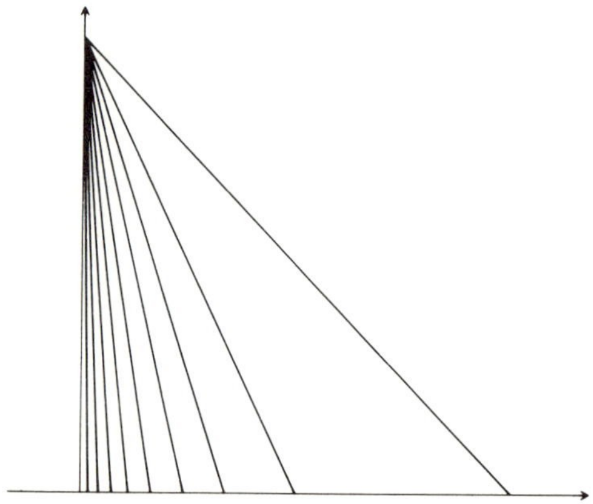

Fig. 3.2

3. The set Y of Theorem 3.4 is not l.p.c. at any point of U.

We will only prove Example 1. Any neighborhood of 0 contains infinitely many points $1/n$, but, clearly, any path in P must take a constant value.

3.6 Theorem *If X is l.p.c. and connected, it is path-connected.*

Proof. Suppose X l.p.c. and not path-connected. We will construct a partition. Suppose that $x, y \in X$ cannot be joined by a path. Let U be the set of points which can be joined to x by a path in X. Write $V = X - U$. Then in order to show that (U, V) is a partition, it remains only to show that U and V are open.

Let $u \in U$. Then u has a path-connected neighborhood W in X. If $w \in W$, there is a path in W joining w to u and a further path in X joining u to x. So w can be joined to X; $w \in U$, and thus $W \subset U$. As U contains a neighborhood of each of its points, it is open. Similarly, let $v \in V$ have a path-connected neighborhood N. If $n \in N \cap U$, v can be joined to x by a path via n, a contradiction. Thus $N \subset V$, and V is open, as required. ∎

FURTHER DEVELOPMENTS

The notions of connectedness and path-connectedness refer essentially to points, so are "zero-dimensional properties." They can be generalized to "n-(path)-connectedness" by replacing points by polyhedra of dimension at most n. The case $n = 1$ will be studied below; for higher n, you can refer to any book on algebraic topology, e.g. Hocking and Young, Maunder or Spanier.

One can also define local n-(path)-connectedness so that analogues of Lemma 3.5 and Theorem 3.6 hold. For some of this, see Hocking and Young; for a very detailed study see Wilder.

Hocking, J. G. and G. S. Young, *Topology*, Addison-Wesley, Reading, Mass., 1961.

Maunder, C. R. F., *Algebraic Topology*, Van Nostrand, Princeton, 1970.

Spanier, E. H., *Algebraic Topology*, McGraw-Hill, New York, 1966.

Wilder, R. L., "Topology of Manifolds," *Amer. Math. Soc. Colloq. Publ.* **32**, (1949).

EXERCISES AND PROBLEMS

1. Determine which of the following spaces are connected and which are path-connected:
   a) The subsets of **C** defined respectively by

      i) $|z| < 1$ ii) $|z| = 1$
      iii) $|z| > 1$ iv) $|z| \geqslant 1$
      v) $|z| \neq 1$ vi) $|z| \leqslant 1$.

b) The subsets of $\mathbf{R}^3$ defined respectively by

i) $x_1^2 + x_2^2 + x_3^2 = 1$ ii) $x_1^2 + x_2^2 - x_3^2 = 1$

iii) $x_1^2 - x_2^2 - x_3^2 = 1$ iv) $-x_1^2 - x_2^2 - x_3^2 = 1$.

c) The sets of real 2×2 matrices which are

i) arbitrary ii) nonsingular iii) singular

iv) symmetric v) orthogonal.

d) The square $\{(x_1, x_2) \in \mathbf{R}^2 : 0 \leqslant x_1 \leqslant 1, 0 \leqslant x_2 \leqslant 1\}$ with the following vertical intervals *removed*:

$$\left.\begin{array}{ll} x = 2^{-2k}, & 0 \leqslant y \leqslant \frac{2}{3} \\ x = 2^{-2k-1}, & \frac{1}{3} \leqslant y \leqslant 1 \end{array}\right\} \text{ for all integers } k \geqslant 1.$$

e) The same as (d), but starting with $0 < x_1 < 1, 0 < x_2 < 1$.

2. Show that if $f: X \to Y$ is a continuous surjective map, then

if X is connected, then Y is connected, and

if X is path-connected, then Y is path-connected.

Would you expect the corresponding result to hold for local path-connectedness? Describe in outline a proof or counterexample.

3. Use Lemma 3.5(iii) to show that S^1 is l.p.c.

4. Prove the facts asserted about the examples.

5. Prove the following: If A and B are connected subsets of $\mathbf{R}^n$, and $A \cap B$ is nonempty, then $A \cup B$ is connected. Does the result remain true if "connected" is replaced by "path-connected" throughout?

6. Show that if A is a connected nonempty subset of $\mathbf{R}^m$, and B a connected nonempty subset of $\mathbf{R}^n$, then $A \times B$ is a connected nonempty subset of $\mathbf{R}^{m+n}$. Prove also the converse.

7. Show that a subset of $\mathbf{R}$ is connected if and only if it is an interval (closed, half-open, or open).

8. Give an example of a path-connected subset X of some euclidean space such that there is no continuous surjective map $I \to X$.

9. X is a connected subset of $\mathbf{R}^n$, and $X \subseteq Y \subseteq Cl(X)$. Prove that Y is connected. Deduce that the space of Theorem 3.4 is connected.

10. Show that if X and Y are homeomorphic, and $x \in X$, then for some $y \in Y$, $X - \{x\}$ is homeomorphic to $Y - \{y\}$. Deduce that no two of $\mathbf{R}^1$, S^1 and $\mathbf{R}^2$ are homeomorphic.

11. Show that if U is a connected open subset of $\mathbf{R}^n$, any two distinct points of U may be joined by a path in U made up of a finite number of straight line segments. [*Hint:* Use the method of proof of Theorem 3.6.] Deduce that the path may be taken to be an embedding $I \to U$. Show also that the two points may be joined by a path $f: I \to U$ such that f has continuous first-order partial derivatives. [*Hint:* Use the path just constructed, and "round off" the corners.]

12. Let $X = X_1 \cup X_2$, where X_1 and X_2 are both closed, l.p.c. subsets of X. Show that X also is l.p.c. What if X_1 is closed and X_2 is open?

4
MORE ABOUT CONNECTION

INTRODUCTION

The characteristic method of procedure of algebraic topology is to start from a relatively straightforward geometrical idea, and then build apparatus on this, using algebraic methods. In this chapter we will develop algebra on the basis of the ideas of connection and path-connection.

THE GROUP $H^0(X)$

Definition *$H^0(X)$ is the set of continuous maps $X \to \mathbf{Z}$.*

The reason for the zero symbol in the notation is that H^0 is the first of a sequence of analogous constructions, based on various dimensions. H^0 is the zero-dimensional case. We will define H^1 in the next chapter.

4.1 Lemma *If $f, g: X \to \mathbf{Z}$ are continuous maps, define $-f: X \to \mathbf{Z}$ and $(f+g): X \to \mathbf{Z}$ by*

$$(-f)(x) = -f(x), \qquad (f+g)(x) = f(x) + g(x), \qquad (x \in X).$$

Then $-f$ and $f+g$ are continuous, and $H^0(X)$ is a group with respect to these operations.

Continuity of $f+g$ was proved on p. 7. We leave the other assertions as exercises. ■

Of course, $\mathbf{Z}$ is discrete in the sense that each subset is open. In this, it resembles the set $\{0, 1\}$ which we used in Chapter 3. We note that X is connected if and only if each map $X \to \mathbf{Z}$ is constant. For, if X is not connected, there is a nonconstant map of X to $\{0, 1\} \subset \mathbf{Z}$. Conversely, if there is a nonconstant map $X \xrightarrow{f} \mathbf{Z}$ with $n \in f(X)$, we define $r: \mathbf{Z} \to \{0, 1\}$ by

$$r(n) = 0,$$
$$r(\mathbf{Z} - \{n\}) = 1;$$

then r is continuous and $r \circ f$ maps X onto $\{0, 1\}$.

The idea of connection has thus given us an abelian group, defined for each space X. This builds up further as follows.

Definition *Let $f: X \to Y$ be a continuous map. We define*

$$f^*: H^0(Y) \to H^0(X)$$

by

$$f^*(g) = g \circ f.$$

4.2 Lemma *f^* is a homomorphism. If 1 is the identity map of X, 1^* is the identity of $H^0(X)$. If $f: X \to Y$ and $g: Y \to Z$, then $(g \circ f)^* = f^* \circ g^*$.*

Proof

$$\begin{aligned} f^*(g_1 + g_2)(x) &= (g_1 + g_2)(f(x)) \\ &= g_1(f(x)) + g_2(f(x)) \\ &= f^*g_1(x) + f^*g_2(x) \\ &= (f^*g_1 + f^*g_2)(x), \end{aligned}$$

so we have a homomorphism. The second statement is clear; so is the third on account of

$$(g \circ f)^*h = h \circ (g \circ f) = (h \circ g) \circ f = f^*(h \circ g) = (f^* \circ g^*)h. \blacksquare$$

These properties seem trivial; indeed, in this case, they are. However, it is worth while to list these properties. We will need them later, and also their analogues in other situations where the proofs are less trivial, so the reader will do well to familiarize himself with them here.

THE SET $\pi_0(X)$

We now investigate path-connection again. We define a binary relation $\sim$ on points of the space X by letting $x \sim y$ when there is a path in X joining x to y. The following property was used implicitly in the previous chapter (where?); we now give a formal proof.

4.3 Lemma $\sim$ *is an equivalence relation on X.*

Proof. Define $f_x: I \to X$ by $f_x(t) = x$ for all $t \in I$. Evidently, f_x is continuous. Since $f_x(0) = f_x(1) = x$, this demonstrates the reflexive property $x \sim x$.

Now define $R: I \to I$ by $R(t) = 1 - t$. Then R is continuous. If $f: I \to X$ is a path joining x to y, $f(0) = x$, $f(1) = y$, then $f \circ R: I \to X$ joins y to x, and is continuous, by C2 of Chapter 1. Thus $\sim$ is symmetric.

Finally, let $f: I \to X$ be a path joining x to y and $g: I \to X$ a path joining y to z, so that

$$f(0) = x, \qquad f(1) = g(0) = y, \qquad \text{and} \qquad g(1) = z.$$

We define a function $h: I \to X$ by

$$\left.\begin{aligned} h(t) &= f(2t) \\ &= g(2t-1) \end{aligned}\right\} \begin{aligned} &\text{if } 0 \leqslant t \leqslant \tfrac{1}{2}, \\ &\text{if } \tfrac{1}{2} \leqslant t \leqslant 1. \end{aligned}$$

The two definitions agree at $t = \frac{1}{2}$ since $f(1) = g(0) = y$; evidently $h(0) = x$, $h(1) = z$. Thus, if h is continuous, it is a path joining x to z, so $x \sim z$, which demonstrates transitivity of $\sim$.

Now f and g are continuous, so clearly h is continuous on each of the closed intervals $[0, \frac{1}{2}]$ and $[\frac{1}{2}, 1]$. By Theorem 1.7, h is continuous. ■

We write $\pi_0(X)$ for the set of equivalence classes of points of X under the relation $\sim$. The equivalence classes themselves are called *path-components* of X. Evidently, X is path-connected if and only if $\pi_0(X)$ contains only one element. We will need the canonical function $X \xrightarrow{p} \pi_0(X)$ which takes each point to its equivalence class. Moreover, in analogy with Lemma 4.2 we have induced mappings. Thus let $f: X \to Y$ be a continuous map of topological spaces. Then if $x \sim x'$ as points of X, we have a path $p: I \to X$ joining them. Then $f \circ p: I \to Y$ is a path joining $f(x)$ to $f(x')$, so $f(x) \sim f(x')$ in Y. Hence f maps equivalence classes into equivalence classes, and thus induces a map

$$f_*: \pi_0(X) \to \pi_0(Y).$$

We can characterize this map by the requirement that $f_* \circ p = p \circ f$ in the (commutative) diagram

$$\begin{array}{ccc} X & \xrightarrow{f} & Y \\ \Big\downarrow p & & \Big\downarrow p \\ \pi_0(X) & \xrightarrow[f_*]{} & \pi_0(Y) \end{array}$$

4.4 Lemma *If f is the identity map of X, f_* is the identity map of $\pi_0(X)$. If $f: X \to Y$ and $g: Y \to Z$, then $(g_* \circ f)_* = g_* \circ f_*$.*

We leave the proof of this as an exercise to the reader. ■

In our program of algebraizing the results of Chapter 3, we next come to the first real result: Theorem 3.4. In our present context, this states that if $\pi_0(X)$ contains only one element, then $H^0(X)$ contains only constant maps. It is reasonable to expect that the same argument will provide information in general which permits us to deduce facts about $H^0(X)$ from facts about $\pi_0(X)$: this we will now obtain.

4.5 Theorem *Let $f: X \to \mathbf{Z}$ belong to $H^0(X)$. Then, if $x \sim x'$ as points of X, $f(x) = f(x')$. Thus f factorizes as a function on the set of equivalence classes,*

$$X \to \pi_0(X) \xrightarrow{c(f)} \mathbf{Z}.$$

Write Map $(\pi_0(X), \mathbf{Z})$ for the set of all integer-valued functions on the set $\pi_0(X)$. Then pointwise addition of functions gives Map $(\pi_0(X), \mathbf{Z})$ the structure of an abelian group. The map defined above,

$$c: H^0(X) \to \text{Map}\,(\pi_0(X), \mathbf{Z})$$

is an injective homomorphism of abelian groups.

Proof. If $x \sim x'$, we can find a path $p: I \to X$ joining x to x'. Then $f \circ p: I \to \mathbf{Z}$ is continuous: since I is connected by Theorem 3.3, it is constant. Thus

$$f(p(0)) = f(p(1)),$$

that is,

$$f(x) = f(x').$$

Indeed, we can alternatively argue that the relation $\sim$ is trivial on $\mathbf{Z}$, so that $\pi_0(\mathbf{Z}) = \mathbf{Z}$. Then $c(f) = \pi_0(f)$.

We leave the proof that Map $(\pi_0(X), \mathbf{Z})$ is an abelian group as an exercise (cf. Chapter 2, H2).

Let ξ be the equivalence class of a general point $x \in X$. That c is a homomorphism follows from the computation (which uses the definition of c and of addition in the two groups)

$$\begin{aligned} c(f_1 + f_2)(\xi) &= (f_1 + f_2)(x) = f_1(x) + f_2(x) \\ &= c(f_1)(\xi) + c(f_2)(\xi) = (c(f_1) + c(f_2))(\xi), \end{aligned}$$

implying $c(f_1 + f_2) = c(f_1) + c(f_2)$.

Finally, if $p: X \to \pi_0(X)$ denotes the projection, then by definition of c, $f = c(f) \circ p$. Clearly, then, if $c(f) = c(g)$, we have

$$f = c(f) \circ p = c(g) \circ p = g. \;\blacksquare$$

Theorem 4.5 contains the first assertion of Theorem 3.4, for if X is path-connected, $\pi_0(X)$ has only one element, so all the maps $\pi_0(X) \to \mathbf{Z}$ are constant. Let us investigate the second assertion. Here we had a space $Y = U \cup V$ which was connected, but not path-connected, although the subsets U and V

were. Clearly, then $\pi_0(Y) = \{U, V\}$. The image of c contains only the constant maps $\pi_0(Y) \to \mathbf{Z}$; thus c is clearly not surjective. However, just as it seemed reasonable before that connectedness should imply path-connectedness for "nice" spaces, so here one can expect c to be surjective for sufficiently well-behaved X. In fact, the same condition as before (l.p.c.) is the one we need. For logical development, we first need a lemma, which was contained in Theorem 3.6, but not there made explicit.

4.6 Lemma *Let X be l.p.c. Then its path-components are open in X.*

Proof. Let ξ be a path-component, $x \in \xi$. Since X is l.p.c. at x, x has a path-connected neighborhood W. All points of W are joinable to x, so $W \subset \xi$. As ξ contains a neighborhood of each of its points, it is open. ■

4.7 Theorem *If X is l.p.c., then*

$$c\colon H^0(X) \to \mathrm{Map}\,(\pi_0(X), \mathbf{Z})$$

is an isomorphism.

Proof. We already know that c is an injective homomorphism; it remains to show, then, that c is surjective. Now c was characterized by $f = c(f) \circ p$. Thus a map $F\colon \pi_0(X) \to \mathbf{Z}$ is in the image of c if and only if

$$f = F \circ p\colon X \to \mathbf{Z}$$

is continuous. Now for each $n \in \mathbf{Z}$, $f^{-1}\{n\}$ is the union of the path-components ξ such that $F(\xi) = n$. By Lemma 4.6, since x is l.p.c., these are open, hence so, by Proposition 1.3, is their union. Thus f is continuous by Theorem 1.1. ■

THE GROUP $H_0(X)$

We now introduce a further element of algebra by remembering the free abelian groups of Chapter 2. Recall that $F(X)$ is the free abelian group on (X, i), where

$$i\colon X \to F(X),$$

and that for any abelian group A, the map

$$\mathrm{Hom}\,(F(X), A) \xrightarrow{i^*} \mathrm{Map}\,(X, A)$$

defined by

$$h \xrightarrow{i^*} h \circ i$$

is bijective.

We will use two of these free abelian groups, namely $F(X)$ and $F(\pi_0(X))$.

Definition $H_0(X) = F(\pi_0(X))$.

By the remark at the end of Chapter 2, the projection $p: X \to \pi_0(X)$ induces a group homomorphism

$$F(p): F(X) \to F(\pi_0(X)) = H_0(X),$$

which is characterized by requiring the diagram

$$\begin{array}{ccc} X & \xrightarrow{p} & \pi_0(X) \\ \downarrow{\scriptstyle i} & & \downarrow{\scriptstyle i} \\ F(X) & \xrightarrow[p_*]{} & H_0(X) \end{array}$$

to commute.

From the universal mapping property of $F(\pi_0(X))$ we deduce in particular that

$$i^*: \text{Hom}\,(H_0(X), \mathbf{Z}) \to \text{Map}\,(\pi_0(X), \mathbf{Z})$$

is an isomorphism. If we combine this with Theorem 4.5 and Theorem 4.7, we find

4.8 Theorem *For any X, $(i^*)^{-1} \circ c = k: H^0(X) \to \text{Hom}\,(H_0(X), \mathbf{Z})$ is an injective homomorphism of abelian groups. If X is l.p.c., it is an isomorphism.* ■

FURTHER DEVELOPMENTS

The trivial properties developed in Lemma 4.3 and Theorem 4.5 constitute respectively the definitions of contravariant and covariant functors. For the theory of categories and functors see Freyd or Mitchell. As our notations imply, H^0, π_0, and H_0 are the first (or rather, zeroth) members of sequences of related functors. In this book we will only continue with H^1, except in Chapter 14, where we define H_1. The customary definitions, however, emphasize algebra to the point of concealing any geometrical significance. The relation $\sim$ will be generalized in the next chapter. In higher dimensions, the functors π_n and H_n are much less closely related. A duality relation analogous to Theorem 4.8 persists, however, for sufficiently well-behaved spaces (for finite simplicial complexes, again, any book on algebraic topology will mention this).

Freyd, P., *Abelian Categories*, Harper and Row, New York, 1964.

Mitchell, B., *Theory of Categories*, Academic Press, New York, 1965.

EXERCISES AND PROBLEMS

1. Establish properties of homomorphisms of the groups $H_0(X)$ induced by continuous maps, in a similar manner to Lemma 4.3 and Theorem 4.5.
2. a) Let $X = U \cup V$, where U and V are disjoint open subsets of X. Construct a bijective correspondence between $\pi_0(X)$ and $\pi_0(U) \cup \pi_0(V)$, and an isomorphism between $H^0(X)$ and the direct product $H^0(U) \times H^0(V)$.
   b) Find results corresponding to those in (a) for the case when U and V are both closed, but have a single common point.
   c) Find the corresponding results when $X = \bigcup_\alpha U_\alpha$ is the disjoint union of an arbitrary family of open subsets U_α.
3. Let $X = Y_1 \times Y_2$. Is there in general a bijective correspondence between $\pi_0(X)$ and $\pi_0(Y_1) \times \pi_0(Y_2)$? Is it true that $H^0(X)$ and $H^0(Y_1) \times H^0(Y_2)$ are in general isomorphic? In each case, give a proof or counterexample.
4. Decide which of the following assertions, about a continuous map $f: X \to Y$, and $f_*: \pi_0(X) \to \pi_0(Y)$, $f^*: H^0(Y) \to H^0(X)$, are true in general. In each case give a proof or counterexample.
   a) If f is surjective, f_* is surjective.
   b) If f is injective, f_* is injective.
   c) If f is bijective, f_* is bijective.
   d) If f is surjective, f^* is injective.
   e) If f is injective, f^* is surjective.
5. Let X be a subset of a euclidean space, Y its closure, $i: X \subset Y$. Show that i^* is injective. Give an example of an X for which i_* is neither injective nor surjective.
6. Find a connected bounded closed subset of $\mathbf{R}^3$ which has an infinite number of path components.
7. Let

$$X = \{0\} \cup \{\tfrac{1}{n} : n \in \mathbf{Z}, n > 0\}.$$

   Describe the image of the homomorphism c.
8. a) Let X be a topological space, x a point of X. The *component*, C_x of X containing x, is defined as the union of all the connected subsets of X which contain x. Show that $x \in C_x$, that C_x is connected, and that C_x is closed in X; also, that C_x contains the path-component of x in X.
   b) Show that the relation between points of X defined by

$$y \sim x \qquad \text{when} \qquad y \in C_x$$

   is an equivalence relation (the equivalence classes are the components of X). Show that when X is l.p.c., its components are the same as the path-components.

*9. Give $\pi_0(X)$ a topology by defining a subset S to be open when $p^{-1}(S)$ is open in X. Verify that this is a topology, and makes p a continuous map. Show that

$$p^*: H^0(\pi_0(X)) \to H^0(X)$$

is an isomorphism.

10. Show that for any X and $\alpha \in H^0(X)$, if $n.\alpha$ $(n \geqslant 1)$ is zero, then $\alpha = 0$.
11. Let X be compact and l.p.c. Show that $\pi_0(X)$ is a finite set.
12. Show that if X and Y are l.p.c., then so is $X \times Y$.

5
DEFINITION OF HOMOTOPY

INTRODUCTION

We now generalize the construction of $\pi_0(X)$ in the preceding chapter. This was obtained by means of an equivalence relation on the points of X. Now if $\{0\}$ is a set with one element, a map $\{0\} \to X$ determines a point of X (its image), and can be identified with that point. We generalize by replacing $\{0\}$ by an arbitrary space. Naturally, this makes the objects under discussion harder to visualize. As a compensating advantage, however, we find that the elementary properties are no harder to establish than in the special case, but by their generality are a good deal more useful, and even lead to some rethinking of our object of study.

DEFINITION OF HOMOTOPY

Two maps $f, g: Y \to X$ are called *homotopic* if there exists a continuous map $F: Y \times I \to X$ such that for all $y \in Y$, we have

$$\left.\begin{aligned} F(y, 0) &= f(y), \\ F(y, 1) &= g(y). \end{aligned}\right\}$$

When this is so, we say that F is a *homotopy* between f and g, and write $F: f \simeq g$. For example, if f is homotopic to a constant map, it is said to be *nullhomotopic*. Sometimes it is convenient to use a different notation for F, and write, for $y \in Y$, $t \in I$,

$$f_t(y) = F(y, t).$$

In this notation the homotopy is less clear, but it is perhaps easier to regard the (continuous) maps $f_t: Y \to X$ as a continuous deformation of $f = f_0$ into $f_1 = g$.

To justify our assertion that f_t is continuous, we write it as a composition

$$Y \xrightarrow{i_t} Y \times I \xrightarrow{F} X,$$

where i_t is the map whose components are the identity map of Y and the constant map with image t. Since these are continuous, so (by C4 of Chapter 1) is i_t, and hence also $f_t = F \circ i_t$.

5.1 Lemma *Homotopy is an equivalence relation between maps $Y \to X$.*

Proof

$f \simeq f$. We use the "constant" homotopy $f_t = f$; more precisely, $F\colon Y \times I \to X$ is defined by $F(y, t) = f(y)$ for all t. This is continuous since $F = f \circ p_1$.

$f \simeq g \Rightarrow g \simeq f$. Let $F\colon f \simeq g$. We define the reflection map of $Y \times I$, R_Y: $Y \times I \to Y \times I$, by

$$R_Y(y, t) = (y, 1 - t).$$

This is continuous, since its component maps p_1 and $R \circ p_2$ are—R as a function of the real variable t. Now $F \circ R_Y\colon g \simeq f$ is continuous since F and R_Y are.

$f \simeq g$ and $g \simeq h \Rightarrow f \simeq h$. Let $F\colon f \simeq g$ and $G\colon g \simeq h$. We must modify the "composition of two paths" construction in Lemma 4.4. We define a map

$$F * G\colon Y \times I \to X$$

by

$$F * G(y, t) = \begin{cases} F(y, 2t), & 0 \leqslant t \leqslant \frac{1}{2}, \\ G(y, 2t - 1), & \frac{1}{2} \leqslant t \leqslant 1. \end{cases}$$

First note that the function has been defined twice at points (y, t) with $t = \frac{1}{2}$. But the two values agree since

$$F(y, 1) = g(y) = G(y, 0).$$

Clearly $F * G$ is continuous on $Y \times [0, \frac{1}{2}]$ and on $Y \times [\frac{1}{2}, 1]$; by Theorem 1.7, it is continuous. ■

Equivalence classes for the relation of homotopy are called *homotopy classes.* We write $[Y, X]$ for the set of homotopy classes of maps $f\colon Y \to X$, and $[f]$ for the equivalence class containing the map f. The above raises the problem of what sort of set $[Y, X]$ is, but gives no hint how to compute it. A little progress toward computation will, however, be given below. Here we observe that we have collected together all maps which can be deformed into one another. This suggests that we have rejected the "continuous part" of the space of all maps $Y \to X$, to concentrate on the "discrete part." From this, it is reasonable to suppose (as is, in fact, the case) that homotopy classes will be more amenable to algebraic manipulations than are maps themselves. We will provide some justification for this view here, and also in Chapter 8.

The following lemma is trivial, but shows the way to manipulation of homotopy sets.

5.2 Lemma *Let*

$$h: Z \to Y, \qquad g_0, g_1: Y \to X, \qquad \text{and} \qquad f: X \to W$$

be continuous maps. If $g_0 \simeq g_1$, *then*

$$g_0 \circ h \simeq g_1 \circ h \qquad \text{and} \qquad f \circ g_0 \simeq f \circ g_1.$$

Proof. Let $G: Y \times I \to X$ be a homotopy of g_0 to g_1. Then

$$f \circ G: Y \times I \to X$$

is a homotopy of $f \circ g_0$ to $f \circ g_1$. For the other part, define $h \times 1: Z \times I \to Y \times I$ by

$$(h \times 1)(z, t) = \big(h(z), t\big).$$

This has continuous components

$$p_1 \circ (h \times 1) = h \circ p_1 \qquad \text{and} \qquad p_2 \circ (h \times 1) = p_2,$$

hence is continuous. And

$$G \circ (h \times 1): Z \times I \to X$$

is a homotopy of $g_0 \circ h$ to $g_1 \circ h$. ■

In the alternative notation, we could have written our homotopies more simply as $f \circ g_t$ and $g_t \circ h$. However, this would have made it harder to prove their continuity as functions on the appropriate product spaces.

Corollary *If* $g_0 \simeq g_1: Y \to X$ *and* $f_0 \simeq f_1: X \to W$, *then*

$$f_0 \circ g_0 \simeq f_1 \circ g_1: Y \to W.$$

For, using both parts of the lemma, we have

$$f_0 \circ g_0 \simeq f_0 \circ g_1 \simeq f_1 \circ g_1,$$

and the result follows from transitivity (Lemma 5.1). Alternatively we could use the homotopy $f_t \circ g_t$. ■

This corollary shows that the homotopy class $[f \circ g]$ depends only on the homotopy classes $[f]$ and $[g]$. We can thus define unambiguously composition of homotopy classes by

$$[f] \circ [g] = [f \circ g].$$

Taking composites thus defines a map

$$[X, W] \times [Y, X] \to [Y, W].$$

Recall that 1_X denotes the identity map of the space X.

5.3 Lemma *Let $h: Z \to Y$, $g: Y \to X$ and $f: X \to W$. Then*

$$[g] \circ [1_Y] = [g] = [1_X] \circ [g],$$
$$([f] \circ [g]) \circ [h] = [f] \circ ([g] \circ [h]). \blacksquare$$

These results are immediate since the composites of the indicated representative maps are already equal. They do, however, give an adequate formalization of the more trivial properties of the homotopy relation.

HOMOTOPY EQUIVALENCE

We now define $f: Y \to X$ to be a *homotopy equivalence*, with homotopy inverse e, if $[f] \circ [e] = [1_X]$, $[e] \circ [f] = [1_Y]$, or equivalently, $f \circ e \simeq 1_X$ and $e \circ f \simeq 1_Y$. When this condition is satisfied, the maps $[Z, Y] \leftrightarrow [Z, X]$ defined by

$$[h] \to [f] \circ [h],$$
$$[e] \circ [k] \leftarrow [k]$$

are bijective (this follows from Lemma 5.3) for any space Z; similarly there are bijections between $[Y, W]$ and $[X, W]$. Thus in the problem of homotopy classification of maps, we can indeed regard the spaces X and Y as equivalent: we call them *homotopy equivalent*.

Example Let Y be the sphere $S^{p-1} \subset \mathbf{R}^p \subset \mathbf{R}^{p+q}$, and X the subset of $\mathbf{R}^{p+q}$ of points *not* lying on the plane $x_1 = \cdots = x_p = 0$. Then the inclusion $f: Y \to X$ is a homotopy equivalence.

For define $e: X \to Y$ by

$$e(x_1, \ldots, x_{p+q}) = (\lambda x_1, \ldots, \lambda x_p, 0, \ldots, 0),$$

where $\lambda = (x_1^2 + \cdots + x_p^2)^{-1/2}$; then $e \circ f = 1_Y$, so it remains to construct a homotopy $H: H \times I \to X$ of $f \circ e$ to 1_X. We find that

$$H((x_1, \ldots, x_{p+q}), t) = (\lambda^{1-t}x_1, \ldots, \lambda^{1-t}x_p, tx_{p+1}, \ldots, tx_{p+q})$$

will do this.

More generally, suppose merely that $[f] \circ [e] = [1_X]$. In this case we say that Y *dominates* X. The reader can verify that the map $[Z, Y] \to [Z, X]$ above is surjective, and $[Z, X] \to [Z, Y]$ is injective (Exercise 16).

A particularly important example of homotopy equivalence occurs when X has only one point x. If Y is homotopy equivalent to a point, Y is called *contractible.* Since our above constructions generalize those of the preceding chapter, we have $[\{x\}, Z] = \pi_0(Z)$, while evidently, $[Z, \{x\}]$ has only one element. It follows from the above remarks that for any contractible space Y, $[Z, Y]$ has only one element whereas $[Y, Z]$ corresponds bijectively to $\pi_0(Z)$.

HOMOTOPY SETS; THE GROUPS $H^1(X)$

Our notation $[X, Y]$ emphasizes the symmetrical dependence of the homotopy set on both spaces. However, it is more convenient for calculations to fix one of the spaces, and regard the homotopy set as depending on the other. When we do this, we introduce a new notation for compositions, and write

$$[f] \circ [g] = f_*[g] = g^*[f].$$

We can then rewrite Lemma 5.3 as

5.4 Lemma *Let $h: Z \to Y$, $g: Y \to X$, and let W be another space. Then we have a map*

$$g^*: [X, W] \to [Y, W];$$

1_X^* *is the identity map of $[X, W]$, and $h^* \circ g^* = (g \circ h)^*$. We also have a map*

$$g_*: [W, Y] \to [W, X];$$

1_{Y*} *is the identity map of $[W, Y]$, and $g_* \circ h_* = (g \circ h)_*$. Moreover, if $f: A \to B$, then*

$$f_* \circ g^* = g^* \circ f_*: [X, A] \to [Y, B]. \blacksquare$$

Observe that Lemma 4.3 is a special case of the first assertion (with $W = \mathbf{Z}$), and Theorem 4.5 of the second (with W a point). It is now time to move on to the next special case of real interest: when W is a circle.

Recall that $S^1 = \{z \in \mathbf{C}: |z| = 1\}$. It follows from standard properties of complex numbers that S^1 is a group, that the multiplication map $m: S^1 \times S^1 \to S^1$ is continuous, and that the inversion map $S^1 \to S^1$ (that is $z \to z^{-1}$) is continuous. A topological space which is also a group and has these properties is called a *topological group.*

5.5 Lemma *For any space X, pointwise multiplication endows the set of continuous maps $X \to S^1$ with the structure of an abelian group. It is compatible with homotopy, thus the set $[X, S^1]$ acquires the structure of a group. If $f: Y \to X$ is continuous, $f^*: [X, S^1] \to [Y, S^1]$ is a homomorphism.*

Proof. Let $c, d \colon X \to S^1$ be continuous maps. The pointwise product $c \cdot d$ can be defined as the composite

$$X \xrightarrow{(c,d)} S^1 \times S^1 \xrightarrow{m} S^1,$$

where (c, d) denotes the (continuous) map with components c and d, and so is continuous. Similarly, using the continuity of inversion, we see that the pointwise inverse c^{-1} is continuous. Clearly the constant map e to 1 acts as unit for this multiplication, c^{-1} is inverse to c, and products are commutative and associative.

Now let $C \colon c \simeq c'$ and $D \colon d \simeq d'$. Then the pointwise product of C and D is continuous (as above), and

$$C \cdot D \colon c \cdot d \simeq c' \cdot d'.$$

Thus $[c \cdot d]$ depends only on $[c]$ and $[d]$, and we may denote it by $[c] \cdot [d]$. Now $[X, S^1]$ inherits an abelian group structure since

$$[e] \cdot [c] = [e \cdot c] = [c],$$
$$[c^{-1}] \cdot [c] = [c^{-1} \cdot c][e]$$
$$[c] \cdot [d] = [c \cdot d] = [d][c]$$

and

$$([c_1] \cdot [c_2]) \cdot [c_3] = [c_1 \cdot c_2 \cdot c_3] = [c_1] \cdot ([c_2] \cdot [c_3]).$$

Finally we must prove $f^*([c] \cdot [d]) = f^*[c] \cdot f^*[d]$. In fact, we prove

$$(c \cdot d) \circ f = (c \circ f) \cdot (d \circ f).$$

For on evaluating at $y \in Y$, the right-hand side gives

$$\begin{aligned} (c \circ f) \cdot (d \circ f)(y) &= (c \circ f)(y) \cdot (d \circ f)(y) \\ &= c(f(y)) \cdot d(f(y)) \\ &= (c \cdot d)(f(y)) \\ &= (c \cdot d) \circ f(y). \ \blacksquare \end{aligned}$$

We will always in future denote $[X, S^1]$ with the group structure described above by $H^1(X)$. The notational similarity to $H^0(X)$ is deliberate, and a direct relation between these will be established in Chapter 8.

FURTHER DEVELOPMENTS

Most of the formal developments of this chapter belong properly speaking to category theory. The axioms describing categories are essentially those verified in Lemma 5.3 for homotopy sets. The calculation of homotopy sets is a highly developed subject; see, for example, Hu, Toda.

Hu, S. T., *Homotopy Theory*, Academic Press, New York, 1959.

Toda, H., "Composition Methods in Homotopy Groups of Spheres," *Ann. of Math. Studies*, **49**, Princeton, (1962).

EXERCISES AND PROBLEMS

1. a) Give details of the proof that if X and Y are homotopy equivalent, then for any W there exists a bijection between $[Y, W]$ and $[X, W]$.
   b) Verify the assertion that if Y dominates X, then for any Z there exist a surjection $[Z, Y] \to [Z, X]$ and an injection $[Z, X] \to [Z, Y]$; also the corresponding statements about maps into Z.
2. Show that any convex subset of euclidean space is contractible.
3. Show that homotopy equivalence is an equivalence relation.
4. Show that the homotopy indicated after Lemma 5.2 by $h_t = f_t \circ g_t$ is in fact continuous.
5. Show that any nonempty space dominates a point.
6. Find spaces S, T such that for any X there is just one continuous map $X \to S$ and just one continuous map $T \to X$.
7. Show that for any spaces W, X, Y there is a bijection of $[W, X \times Y]$ on $[W, X] \times [W, Y]$.
8. Let X, Y be subpsaces of some euclidean space E; define $X \amalg Y$ to be the subspace $X \times 0 \cup Y \times 1$ of $E \times I$. Show that for any W there is a bijection of $[X \amalg Y, W]$ on $[X, W] \times [Y, W]$.
9. Let X and Y be homotopy equivalent, and W any space. Show that the following pairs are homotopy equivalent.
   a) $W \times X$ and $W \times Y$ b) $X \times X$ and $X \times Y$ c) $W \amalg X$ and $W \amalg Y$
10. Show that $\mathbf{R}^{n+1} - \{0\}$ is homotopy equivalent to S^n.
11. Show that $\mathbf{C} - \{0, 1\}$, $S^1 \times 1 \cup 1 \times S^1$, and $S^1 \times S^1 - \{(1, 1)\}$ are all homotopy equivalent.
12. Let $X = \{(p, q) \in S^n \times S^n : p \neq -q\}$. Show that the map $p \mapsto (p, p)$ from S^n to X is a homotopy equivalence.
13. Show that $\{(a, b, c) \in \mathbf{R}^3 : b^2 > ac\}$ is homotopy equivalent to a circle. Interpret this result by considering the roots of the equation $ax^2 + 2bxy + cy^2 = 0$.
14. Show that
$$X = \{(a, b, c, d) \in \mathbf{R}^4 : ax^3 + 3bx^2y + 3cxy^2 + dy^3 = 0 \text{ defines three distinct lines through the origin}\}$$
dominates S^1. [*Hint:* Obtain a map $X \to S^1$ by letting the lines make angles $\theta_1, \theta_2, \theta_3$ with the x-axis and considering $\theta_1 + \theta_2 + \theta_3$ (mod π).] [*Further hint:* X is defined by $6abcd + 3b^2c^2 > a^2d^2 + 4(ac^3 + b^3d)$, but this is not relevant.]

15. Let X be a space, $f: S^1 \to X$. Show that f is nullhomotopic (homotopic to a constant map) if and only if there is a continuous map $g: D^2 \to X$ with $g|S^1 = f$. [*Hint:* If $F: c \simeq f$ with c constant, define $g(rx) = F(x, r)$ for $x \in S^1$, $r \in I$.]

*16. i) Prove that if multiplication (of quaternions) is defined by

$$(x_0, x_1, x_2, x_3).(y_0, y_1, y_2, y_3) = (x_0y_0 - x_1y_1 - x_2y_2 - x_3y_3,$$
$$x_0y_1 + x_1y_0 + x_2y_3 - x_3y_2, x_0y_2 - x_1y_3 + x_2y_0 + x_3y_1,$$
$$x_0y_3 + x_1y_2 - x_2y_1 + x_3y_0),$$

S^3 becomes a (nonabelian) topological group, and for any space X, $[X, S^3]$ is a group.

ii) Prove the same for the groups U_n of $n \times n$ unitary matrices and SU_n of those with determinant $+1$.

iii) Show that the homeomorphism of p. 10 is not an isomorphism of groups, but induces an isomorphism $S^3 \to SU_2$.

6
A STUDY OF A CIRCLE

INTRODUCTION

We begin this chapter by examining the exponential map from $\mathbf{R} \to S^1$, giving full proofs of results sometimes taken as self-evident. This enables us to define the important concept of degree of a map from S^1 to itself. After establishing the basic properties of degree, we apply them to prove the so-called *fundamental theorem of algebra*, and *Brouwer's fixed point theorem in the plane*.

LIFTING MAPS FROM S^1 UP TO R

We define the map $e: \mathbf{R} \to S^1$ by $e(t) = \exp(2\pi i t)$. The usual properties of the exponential function show that e is continuous and surjective, and that it is a homomorphism from the additive group $\mathbf{R}$ to the multiplicative group S^1:

$$e(t + u) = e(t)\, e(u).$$

The kernel of this homomorphism is the subgroup $\mathbf{Z}$ of integers.

We propose to use the map e to study topological properties of the circle. The reader is asked to picture $\mathbf{R}$ as "spread out above the circle," rather like the helix which is the graph of e (Fig. 6.1).

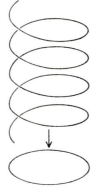

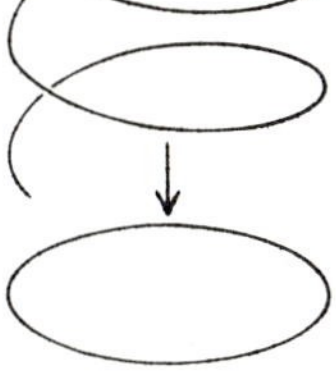

Fig. 6.1

Let X be a topological space, $f: X \to S^1$ a continuous map. We consider the problem: Does there exist a continuous map $\tilde{f}: X \to \mathbf{R}$ with $f = e \circ \tilde{f}$?

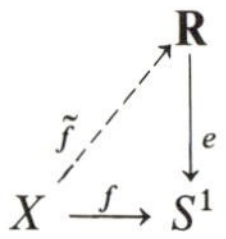

If $\tilde{f}$ can be found, we call it a *lift* of f, and say that f can be *lifted*. The problem of existence of $\tilde{f}$ is known as the lifting problem. We confine ourselves to a few special cases in this chapter; the general case will be considered in the next.

We will need rather detailed topological properties of e below; we begin by collecting them.

6.1 Lemma *The map e': $]0, 1[\to S^1 - \{1\}$ obtained by restricting e is a homeomorphism. Conversely, let B be any subset of $S^1 - \{1\}$, and $A = I \cap e^{-1}(B)$. Then $e^{-1}(B)$ is the union of the sets $A + n$ ($n \in \mathbf{Z}$), each is open in $e^{-1}(B)$, and e induces a homeomorphism of each onto B.*

Proof. It follows from the above that e' is continuous and bijective. It remains to show that e'^{-1} is continuous. For each $x \in S^1 - \{1\}$, choose a closed interval $A \subset\ ]0, 1[$ with $e'^{-1}(x)$ in the interior of A. Since A is compact, by the corollary to Theorem 1.11, e' induces a homeomorphism of A onto $e'(A)$. But $e'(A)$ is clearly a neighborhood of x; it follows that e'^{-1} is continuous at x.

To prove the second part, it suffices to prove it for $B_0 = S^1 - \{1\}$ (the other cases follow). But $e^{-1}(B_0)$ is the disjoint union of the open intervals $]n, n + 1[\ = A_0 + n$ ($n \in \mathbf{Z}$), and by the first part, e induces a homeomorphism of each of these onto B_0. ∎

Remark. The point $1 \in S^1$ plays only a notational role above; an analogous conclusion follows for any proper subset B of S^1.

Corollary *Suppose $f: X \to S^1$ is not surjective. Then f is nullhomotopic.*

For the lemma shows that S^1 with a point deleted is homeomorphic to $]0, 1[$, and so is contractible. ∎

We are now able to prove

6.2 Theorem *Any continuous map $f: I \to S^1$ has a lift $\tilde{f}: I \to \mathbf{R}$, which is unique up to translation by an integer. Thus, if $a_0 \in \mathbf{R}$ with $e(a_0) = f(0)$ is given, there is a unique lift $\tilde{f}$ with $\tilde{f}(0) = a_0$.*

Proof. For each $t \in I$, continuity of f at t implies that we can find an $\varepsilon > 0$ such that $f[t - \varepsilon, t + \varepsilon]$ is a proper subset of S^1. [Strictly speaking, f is only defined on $[t - \varepsilon, t + \varepsilon] \cap I$.] By the Heine–Borel theorem (Lemma 1.10), we can find a finite set of intervals $]t - \varepsilon, t + \varepsilon[$ which cover I. Denote the set of all the endpoints in I of these intervals by

$$0 = t_0 < t_1 < t_2 < \cdots < t_n = 1.$$

Then each $[t_{i-1}, t_i]$ is contained in some interval $[t - \varepsilon, t + \varepsilon]$, so its image by f is a proper subset S_i of S^1.

We will now prove, by induction on i, that there is a unique lift of $f|[0, t_i]$ to a map $\tilde{f}\colon [0, t_i] \to \mathbf{R}$ with $\tilde{f}(0) = a_0$. For $i = 0$ this is trivial; assume it true for $i - 1$. Let $b_i \notin S_i$ (which is a proper subset); let

$$e(c_i) = b_i,$$

and let

$$A_i = e^{-1}(S_i) \cap [c_i, c_i + 1].$$

Then Lemma 6.1 shows that $e^{-1}(S_i)$ is a disjoint union of open sets $A_i + n$, each mapped homeomorphically onto S_i. Let n_i be the integer such that $\tilde{f}(t_{i-1}) \in A_i + n_i$, and let $e_i\colon A_i + n_i \to S_i$ be the homeomorphism induced by e. Then we can define $\tilde{f}$ on $[t_{i-1}, t_i]$ as $e_i^{-1} \circ f$. By Theorem 1.7, this combines with $\tilde{f}$ on $[0, t_{i-1}]$ to give a continuous function on $[0, t_i]$. Clearly, it lifts f. Conversely, any continuous lift maps $[t_{i-1}, t_i]$ to $\bigcup_n (A_i + n)$: since $[t_{i-1}, t_i]$ is connected, it must all be mapped to a single $A_i + n$ (they are disjoint open sets, and so define a partition), and as t_{i-1} is mapped into $A_i + n_i$, so must be all of $[t_{i-1}, t_i]$. Thus the lift is uniquely determined. This completes the induction step, and with it the proof of the theorem. ■

6.3 Lemma *Theorem* 6.2 *continues to hold if* I *is replaced by* I^2 *and* 0 *by* (0, 0).

Proof. This follows the same pattern: we will only indicate the points where it differs from the preceding proof.

Instead of intervals $[t - \varepsilon, t + \varepsilon]$, we use small rectangles with sides parallel to the axes. The Heine–Borel theorem is again applicable. We then use all x-coordinates of vertical edges of rectangles to dissect the first factor I as before, and the y-coordinates of horizontal edges to dissect the second factor $0 = u_0 < u_1 < \cdots < u_m = 1$. Now extend $\tilde{f}$ inductively over the rectangles

$$C_{ij} = \{(t, u) : t_{i-1} \leqslant t \leqslant t_i, u_{j-1} \leqslant u \leqslant u\}$$

ordered by letting C_{ij} precede C_{kl} if $j < l$ or if $j = l$ and $i < k$.

The only further point to be noted is that when we attempt to define $\tilde{f}$

on C_{ij}, having already defined it on all earlier squares, the part of C_{ij} on which $\tilde{f}$ is already defined (i.e. the intersection of C_{ij} with earlier squares) consists of the left-hand side, or of the lower side, or of their union (or, in the case of C_{11}, the lower left-hand corner). In each case, this set is nonempty and connected. Thus its image by $\tilde{f}$, which lies in $\bigcup_n (A_{ij} + n)$, must in fact lie in $A_{ij} + n_{ij}$ for a suitable integer n_{ij}. The rest of the proof is unchanged.

THE DEGREE OF A MAP

We are now ready to study maps of S^1 into itself. Let $f: S^1 \to S^1$; consider the diagram

$$\begin{array}{ccc} I & \overset{g}{\dashrightarrow} & \mathbf{R} \\ \big\downarrow{\scriptstyle e|I} & & \big\downarrow{\scriptstyle e} \\ S^1 & \xrightarrow{f} & S^1 \end{array}$$

By Theorem 6.2, $f \circ (e|I)$ lifts to a map g. As $e(0) = e(1) = 1$, we have

$$e(g(1)) = f(e(1)) = f(e(0)) = e(g(0)),$$

so $g(1) - g(0)$ is an integer. This integer is called the *degree of* f, and denoted by $\deg f$. Any other lift g' is obtained from g by translating by an integer n, so

$$g'(1) - g'(0) = g(1) + n - (g(0) + n) = g(1) - g(0);$$

the degree is independent of the choice of lift. We now show that the notion of degree is adequate to solve all our problems about maps of circles.

6.4 Theorem *Degree defines a group isomorphism*

$$\deg: H^1(S^1) \cong \mathbf{Z}.$$

The following conditions on $f: S^1 \to S^1$ are equivalent:

i) f is nullhomotopic.

ii) f has degree zero.

iii) f has a continuous lift $\tilde{f}: S^1 \to \mathbf{R}$.

Proof. The hardest part of the theorem is to show that homotopic maps have the same degree, so that deg can be regarded as a function on $H^1(S^1)$. We prove this first.

Let $F: S^1 \times I \to S^1$ be a homotopy between f and f'. Consider the diagram

$$\begin{array}{ccc} I \times I & \xrightarrow{G} & \mathbf{R} \\ \Big\downarrow {\scriptstyle (e|I)\times 1} & & \Big\downarrow {\scriptstyle e} \\ S^1 \times I & \xrightarrow{F} & S^1 \end{array}$$

By Lemma 6.3, a lift G can be constructed. We use t, u as coordinates in $I \times I$. Now f is lifted by g, where

$$g(t) = G(t, 0).$$

Thus

$$\deg f = G(1, 0) - G(0, 0).$$

Similarly,

$$\deg f' = G(1, 1) - G(0, 1).$$

Now consider the function of u defined on I by

$$d(u) = G(1, u) - G(0, u).$$

Since $e(G(1, u)) = F(1, u) = e(G(0, u))$, d takes integer values. Also it is continuous. Since I is connected, it follows that d is constant, so

$$\deg f = d(0) = d(1) = \deg f',$$

as required.

Thus we have a well-defined map

$$\deg: H^1(S^1) \to \mathbf{Z}.$$

It is a homomorphism, for let $f, f': S^1 \to S^1$ be continuous, and let $g, g': I \to \mathbf{R}$ lift $f \circ (e|I)$ and $f' \circ (e|I)$ respectively. Then since e is a homomorphism, $g + g'$ lifts $(f \cdot f') \circ (e|I)$. Hence

$$\begin{aligned} \deg (f \cdot f') &= (g + g')(1) - (g + g')(0) \\ &= g(1) - g(0) + g'(1) - g'(0) = \deg f + \deg f'. \end{aligned}$$

To show that deg is surjective, it is now enough to exhibit a map of degree 1. In fact, for any integer n, consider the nth power map

$$p_n: S^1 \to S^1$$

defined by $p_n(z) = z^n$. Since, again, e is a homomorphism, $p_n \circ (e|I)$ is lifted by the map $t \to nt$. Hence p_n has degree $n \cdot 1 - n \cdot 0 = n$.

To show that deg is injective, it suffices to show that it has zero kernel, i.e., that a map of zero degree is nullhomotopic. This is the statement (ii) $\Rightarrow$ (i), so will follow from the last part, which we now establish.

(i) $\Rightarrow$ (ii), since if f is homotopic to f_0, $\deg f = \deg f_0$ by the first part, and a constant map has zero degree (proof as for p_0 above).

(ii) $\Rightarrow$ (iii), for when f has degree zero, we have a lift g of $f \circ (e|I)$ with $g(1) = g(0)$. We now define $\tilde{f}: S^1 \to \mathbf{R}$ lifting f as $g \circ (e|I)^{-1}$. The only possible ambiguity occurs at 1, with $(e|I)^{-1}\{1\} = \{0, 1\}$, but $g(0) = g(1)$. Continuity of $\tilde{f}$ at points other than 1 follows from that of g and of $(e|I)^{-1}$ (shown in Lemma 6.1). Continuity at 1 is shown as in Theorem 1.7, since the map is continuous on each side. We can also divide S^1 into two semicircles, and apply Theorem 1.7 directly.

(iii) $\Rightarrow$ (i), since $\mathbf{R}$ is contractible, so any map $\tilde{f}: S^1 \to \mathbf{R}$ is homotopic to a constant map $\tilde{f}_0$; by Lemma 5.2, $f = e \circ \tilde{f} \simeq e \circ \tilde{f}_0$. ■

APPLICATIONS

We now give an important application of the notion of degree.

6.5 Theorem (*Fundamental Theorem of Algebra*). *Any polynomial equation in* $\mathbf{C}$ *has a root.*

Proof. We will suppose not, and find a contradiction. Write the polynomial as

$$P(z) = \sum_{i=0}^{n} a_i z^i, \qquad \text{with} \qquad a_n \neq 0.$$

Dividing through by a_n, we may suppose $a_n = 1$ without loss of generality. We define a map $F: S^1 \times \mathbf{R}_+ \to S^1$ by

$$F(z, r) = \frac{P(rz)}{|P(rz)|}.$$

This is well defined if P does not vanish, and is clearly continuous. Moreover, if we write $f_r(z) = F(z, r)$, then F provides a homotopy between the maps f_r. Now f_0 is constant, so has zero degree. We will show that for r large enough, f_r has degree n. This will give the required contradiction.

Choose

$$R > \max\left(\sum_{i=0}^{n-1} |a_i|, 1\right).$$

Then for $|z| = 1$,

$$\begin{aligned} \left|\sum_{i=0}^{n-1} a_i(Rz)^i\right| &\leqslant \sum_{i=0}^{n-1} |a_i| R^i \\ &\leqslant R^{n-1} \sum_{i=0}^{n-1} |a_i| \\ &< R^n = |(Rz)^n|. \end{aligned}$$

So

$$\left| \frac{\sum_{i=0}^{n-1} a_i(Rz)^i}{(Rz)^n} \right| < 1.$$

It follows, in particular, that $P(Rz)/(Rz)^n$ has a positive real part. Hence so has

$$\frac{P(Rz)}{(Rz)^n} \left| \frac{(Rz)^n}{P(Rz)} \right| = \frac{f_R(z)}{z^n}.$$

By the corollary to Lemma 6.1, the map

$$z \to \frac{f_R(z)}{z^n}$$

has zero degree. Hence $\deg f_R$ equals the degree of the map $z \to z^n$, which was shown above to be n. The theorem now follows. ∎

Another important application is

6.6 Theorem (*Brouwer's fixed point theorem in the plane*). *Any map $f: D^2 \to D^2$ has a fixed point.*

Proof. Suppose not. Then for $x \in D^2$, x and $f(x)$ are distinct. Define $\phi(x)$ to be the point in which the segment from $f(x)$ to x produced (Fig. 6.2, p. 72) meets S^1. Clearly, ϕ depends continuously on x. Thus $\phi: D^2 \to S^1$ is a map; now $\phi|S^1$ is the identity, with degree 1. On the other hand, D^2 is contractible, so ϕ is homotopic to a constant map; in particular, so is $\phi|S^1$. Hence the degree is zero: a contradiction. ∎

FURTHER DEVELOPMENTS

We will press on with further calculations in the following chapter. The methods employed here are somewhat special to the particular case involved.

Most of the results, however, are susceptible of generalization. For example, Theorem 6.4 generalizes to the construction of a bijection

$$\deg\colon [S^n, S^n] \to \mathbf{Z}.$$

Any book on homotopy theory (e.g. Hu, Maunder or Spanier) will give a proof.

Hu, S. T., *Homotopy Theory*, Academic Press, New York, 1959.

Maunder, C. R. F., *Algebraic Topology*, Van Nostrand, Princeton, 1970.

Spanier, E. H., *Algebraic Topology*, McGraw-Hill, New York, 1966.

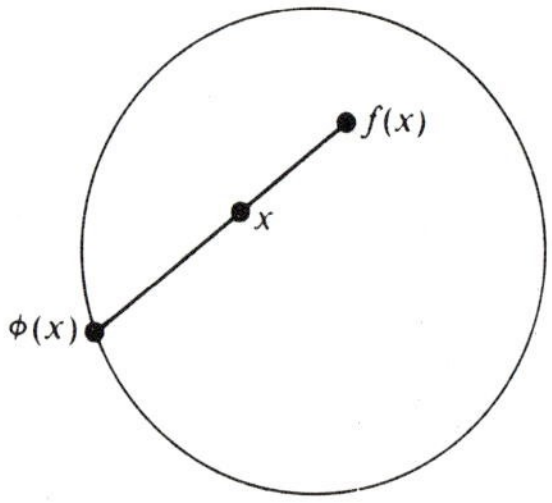

Fig. 6.2

EXERCISES AND PROBLEMS

1. Use the result of Exercise 7, Chapter 1, to generalize the corollary to Lemma 6.1 to S^n.
2. a) Let f and g be maps $S^1 \to S^1$, with $f(1) = g(1) = 1$. Define $f * g\colon S^1 \to S^1$ by

$$f * g(z) = \begin{cases} f(z^2) & \text{if } z = x + iy, y \geqslant 0, \\ g(z^2) & \text{if } z = x + iy, y \leqslant 0. \end{cases}$$

   Show that $f * g$ is well defined and continuous.

   b) Compute each of $\deg(f \cdot g)$, $\deg(f \circ g)$, $\deg(f * g)$ in terms of $\deg f$ and $\deg g$.

   c) Are any two of the maps

$$1, f, g, f \cdot g, f \cdot g^{-1}, f^{-1} \cdot g, f * 1, f * g, f * g^{-1}, g * f, f \circ g,$$
$$f \circ g^{-1}, g \circ f, g \circ f^{-1}$$

   necessarily homotopic? Write down as many explicit homotopies as you can find between these maps.

3. Show that a polynomial equation of degree n over $\mathbf{C}$ has n roots (counted with multiplicities). [*Hint:* Use the remainder theorem.]

4. Let P be a polynomial function on $\mathbf{C}$ which does not vanish on S^1. Show that the number of roots of $P(z) = 0$ with $|z| < 1$ is equal to the degree of the map $f: S^1 \to S^1$ defined by $f(z) = P(z)/|P(z)|$. [*Hint:* Factorize P, and prove the result first when P is linear.]
5. Let S be as in Lemma 6.1. Show that $e^{-1}(S)$ is homeomorphic to $S \times \mathbf{Z}$.
6. Let $f: S^1 \to S^1$ have $f(1) \neq 1$; let $\phi: \mathbf{R} \to \mathbf{R}$ be differentiable and satisfy $e \circ \phi = f \circ e$, and suppose whenever $\phi(x)$ is an integer, $\phi'(x) \neq 0$. Let there be p values of x, $0 < x < 1$, with $\phi(x)$ an integer, $\phi'(x) > 0$, and q values with $\phi'(x) < 0$. Show that $\deg f = p - q$.
7. Show that any map $f: I \to I$ has a fixed point.
8. Show that if $f: S^1 \to S^1$ and $\deg f \neq 1$, then f has a fixed point.

*9. Define a map $f: Y \to X$ to be a *covering map* if for every $x \in X$ we can find a neighborhood N of x in X, a discrete space D (e.g. a finite set or $\mathbf{Z}$) and a homeomorphism $\phi: N \times D \to f^{-1}(N)$ with $f(\phi(n, d)) = n$ for all $n \in N$, $d \in D$. Prove the following.

a) If $f: Y \to X$ is a covering map, $X' \subset X$ and $Y' = f^{-1}(X')$, then $f|Y': Y' \to X'$ is a covering map.

b) If $f_1: Y_1 \to X_1$ and $f_2: Y_2 \to X_2$ are covering maps, then so is
$$f_1 \times f_2: Y_1 \times Y_2 \to X_1 \times X_2.$$

c) If $f: Y \to X$ and $g: Z \to Y$ are covering maps, then so is
$$f \circ g: Z \to X.$$

d) $p_n: S^1 \to S^1$ is a covering map.

e) If $f: Y \to X$ is a covering map, $p: I \to X$, and $y \in Y$ satisfies $f(y) = p(0)$, there is a unique $\tilde{p}: I \to Y$ with $f \circ \tilde{p} = p$ and $\tilde{p}(0) = y$.

f) Let
$$S^1 \vee S^1 = (S^1 \times \{1\}) \cup (\{1\} \times S^1) \stackrel{i}{\subset} S^1 \times S^1,$$
(notation as in Lemma 8.3), and let $f: L \to S^1 \vee S^1$ be the covering induced [as in (a) above] from
$$e \times e: \mathbf{R} \times \mathbf{R} \to S^1 \times S^1.$$
Describe $L \subset \mathbf{R}^2$, and find a map $g: S^1 \to L$ which is not nullhomotopic. Show that
$$f \circ g: S^1 \to S^1 \vee S^1$$
is not nullhomotopic, but $i \circ f \circ g$ is.

10. Show that if $x \in H^1(X)$ and $n \cdot x = 0$, then $x = 0$. [*Hint:* Represent x by $f: X \to S^1$, show that $p_n \circ f \simeq 0$, and lift it).

*11. Let $f: S^1 \to S^1$. Define $f': S^1 \to \mathbf{C}$ (by differentiation) so that
$$\deg f = \frac{1}{2\pi i} \oint_{S^1} \frac{f'(z)}{f(z)} dz.$$

7
LIFTING AND EXTENSION PROBLEMS

INTRODUCTION

We have seen that topology is the study of topological spaces and continuous maps, and that the construction and classification (e.g. by homotopy) of spaces and maps play a large role in the development of the subject. One key problem in constructing maps is factorization.

There are two forms of this.

Lifting problem. Given $f: X \to Y$, $g: Z \to Y$, when does there exist $h: X \to Z$ with $g \circ h = f$?

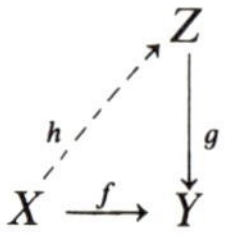

Extension problem. Given $g: B \to A$, $f: B \to C$, when does there exist $h: A \to C$ with $h \circ g = f$?

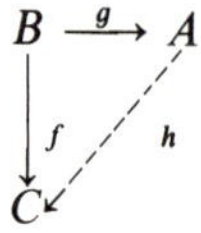

(This is called the extension problem because usually in practice B is a subspace of A, g the inclusion, so h—if it exists—extends the map f defined on B to all of A.)

The method of algebraic topology gives a standard procedure for proving nonexistence of factorizations, which we illustrate with a simple example.

Example

$B = C = S^1$, $A = D^2, f =$ identity, $g =$ inclusion.

There is no extension of $1: S^1 \to S^1$ to a map $D^2 \to S^1$. For if there were, consider

$$\begin{array}{ccc} \mathbf{Z} \cong H^1(B) & \xleftarrow{g^*} & H^1(A) = \{0\} \\ & {\scriptstyle f^* = 1} \nwarrow \quad \nearrow {\scriptstyle h^*} & \\ & \mathbf{Z} \cong H^1(C) & \end{array}$$

We have $1 = f^* = g^*h^* = 0$. But the zero and identity maps $\mathbf{Z} \to \mathbf{Z}$ are not the same: a contradiction.

In general, one can first ask if an h^* exists to solve the algebraic factorization problem:

$$h^* \circ g^* = f^* \qquad \text{(for the lifting problem),}$$

$$g^* \circ h^* = f^* \qquad \text{(for the extension problem).}$$

If we cannot find an h^* satisfying this, there is certainly no continuous map h.

Positive solutions to the problem can only come from direct methods of construction of maps h. We give two such techniques in this chapter. Since direct construction is hard work, these are the most difficult proofs in this book.

THE LIFTING PROBLEM

For the lifting problem, we will only consider the case when $g: Z \to Y$ is $e: \mathbf{R} \to S^1$. We now give two results on this. The first gives a better computational test (especially when π_1 or H_1 is known, see (14.7)). The second will be more useful theoretically.

7.1 Theorem (*Monodromy Theorem*). *Let X be a l.p.c. space, $f: X \to S^1$ a map. Then f has a lift $\tilde{f}: X \to \mathbf{R}$ if and only if for all continuous maps $l: S^1 \to X$, deg $(f \circ l) = 0$.*

Such maps l are called loops (in X).

Proof. First suppose $\tilde{f}$ exists. Then $\tilde{f} \circ l$ is a lift of $f \circ l$; by Theorem 6.4, we have deg $(f \circ l) = 0$. Now suppose, conversely, that this holds for any loop l.

Since (by Theorem 4.7) path components are open in X, if for each path component C we can lift $f|C$ to a map $\tilde{f}|C$, the combined map $\tilde{f}: X \to \mathbf{R}$ will be continuous when all the $\tilde{f}|C$ are. So we can suppose X path-connected.

Now choose $x_0 \in X$, and $a_0 \in \mathbf{R}$ with $e(a_0) = f(x_0)$. For any $x \in X$, choose a path $p: I \to X$ joining x_0 to x. By Theorem 6.2 we can lift $f \circ p$ uniquely to $\widetilde{f \circ p}: I \to \mathbf{R}$, with $\widetilde{f \circ p}(0) = x_0$. Now if $\tilde{f}$ exists, $\tilde{f} \circ p$ lifts $f \circ p$, so equals $\widetilde{f \circ p}$, thus $\tilde{f}(x)$ must equal $\tilde{f}(p(1)) = \widetilde{f \circ p}(1)$. Thus we define $\tilde{f}(x) = \widetilde{f \circ p}(1)$.

We next show that this value is independent of p. For let $q: I \to X$ be another path joining x_0 to x. Then p and q define a loop $l: S^1 \to X$ by

$$l(e(t)) = \begin{cases} p(1 - 2t), & 0 \leqslant t \leqslant \frac{1}{2} \\ q(2t - 1), & \frac{1}{2} \leqslant t \leqslant 1. \end{cases}$$

Note that at $t = \frac{1}{2}$ we have $p(0) = x_0 = q(0)$; also that

$$l(e(0)) = p(1) = x = q(1) = l(e(1)),$$

so the definitions are consistent. Since e^{-1} is continuous (Lemma 6.1), l is continuous on each of the semicircles $\operatorname{Im} z \geqslant 0$, $\operatorname{Im} z \leqslant 0$. That l is continuous now follows by Theorem 1.7. By hypothesis, $\deg(f \circ l) = 0$. But we know a lift of $f \circ l$, namely g, where

$$g(t) = \begin{cases} \widetilde{f \circ p}(1 - 2t), & 0 \leqslant t \leqslant \frac{1}{2} \\ \widetilde{f \circ q}(2t - 1), & \frac{1}{2} \leqslant t \leqslant 1. \end{cases}$$

(Note that these, also, agree at $t = \frac{1}{2}$, having the common value a_0.) Hence $g(0) = g(1)$, that is $\widetilde{f \circ p}(1) = \widetilde{f \circ q}(1)$, as claimed.

Thus $\tilde{f}: X \to \mathbf{R}$ is a well-defined function, with $e \circ \tilde{f} = f$. It remains to show that $\tilde{f}$ is continuous.

Let $x \in X$. Since f is continuous, x has a neighborhood W such that $f(W)$ is a proper subset of S^1. Since X is l.p.c., we can find a path-connected neighborhood $U \subset W$ of x. Set $f(U) = S \subset S^1$. Then by Lemma 6.1 we can write $e^{-1}(S)$ as a disjoint union of translates of some set; denote by A the translate containing $\tilde{f}(x)$. We will show $\tilde{f}(U) \subset A$. It then follows that $\tilde{f}|U$ is the composite of f and the homeomorphism induced by e^{-1} of S on A. So $\tilde{f}|U$ is continuous: as U is a neighborhood of x, $\tilde{f}$ is continuous at x.

To show $\tilde{f}(U) \subset A$, let $y \in U$. Choose the path p from x_0 to y to go first to x, then inside U (which is path-connected) to y. Then $f \circ p$ goes from $f(x_0)$ to $f(x)$, then inside S to $f(y)$. So $\widetilde{f \circ p}$ goes from q_0 to $\tilde{f}(x)$ (by definition of the latter) and then inside $e^{-1}(S)$ to $\tilde{f}(y)$. But a path in $e^{-1}(S)$ which begins in A must lie wholly in A (cf. proof of Theorem 6.2), so $\tilde{f}(y) \in A$, as asserted. ∎

As a corollary of this we have the next result.

7.2 Theorem *Let X and Y be connected and l.p.c., $f: X \times Y \to S^1$ a map, and $(x_0, y_0) \in X \times Y$ a point such that $f|X \times \{y_0\}$ and $f|\{x_0\} \times Y$ lift to maps into* $\mathbf{R}$. *Then f also has a continuous lift.*

Proof. We must show that for any loop $l: S^1 \to X \times Y$, $\deg(f \circ l) = 0$; in fact it follows from the proof of Theorem 7.1 that it is enough to show this for loops with $l(1) = (x_0, y_0)$. Let $l \circ (e|I)$ have components $p: I \to X$, $q: I \to Y$. We must show that if this is lifted to a map $\tilde{g}: I \to \mathbf{R}$, then $\tilde{g}(1) = \tilde{g}(0)$.

Define maps $P: I \times I \to X$, $Q: I \times I \to Y$ by

$$\begin{aligned} P(t, u) &= p\left(\frac{2t}{2 - u}\right), \quad 0 \leqslant t \leqslant 1 - u/2 \\ &= p(1) = x_0, \quad 1 - u/2 \leqslant t \leqslant 1; \\ Q(t, u) &= q(0) = y_0, \quad 0 \leqslant t \leqslant u/2 \\ &= q\left(\frac{2t - u}{2 - u}\right) \quad u/2 \leqslant t \leqslant 1. \end{aligned}$$

Then, for any t, $P(0, u) = x_0 = P(1, u)$ and $Q(0, u) = y_0 = Q(1, u)$. Write

$$G = F \circ (P \times Q): I \times I \to S^1.$$

Then by Lemma 6.3 G has a continuous lift $\tilde{G}: I \times I \to \mathbf{R}$. Since we have $G(0, u) = G(1, u)$, then $\tilde{G}(1, u) - \tilde{G}(0, u)$ is an integer, and so is independent of u.

Putting $u = 0$, we see that this integer is just $\tilde{g}(1) - \tilde{g}(0)$. Now put $u = 1$. We find that for $0 \leqslant t \leqslant \frac{1}{2}$, $(P \times Q)(t, 1)$ describes a loop in $X \times \{y_0\}$, and for $\frac{1}{2} \leqslant t \leqslant 1$, a loop in $\{x_0\} \times Y$. Our hypothesis implies that each of these, composed with F, has degree 0. Thus

$$\tilde{G}(0, 1) = \tilde{G}(\tfrac{1}{2}, 1) = \tilde{G}(1, 1),$$

and our integer is 0. ∎

We now come to our second criterion for the existence of $\tilde{f}$: this is that f be homotopic to a constant. More generally, we have

7.3 Theorem (*Lifting Homotopy Property for e*). *Let X be a space, $\tilde{f}_0: X \to \mathbf{R}$ a map, and $F: X \times I \to S^1$ a homotopy such that for $x \in X$, $F(x, 0) = e(\tilde{f}_0(x))$. Then there is a unique homotopy $\tilde{F}: X \times I \to \mathbf{R}$ with $e \circ \tilde{F} = F$ and $\tilde{F}(x, 0) = \tilde{f}_0(x)$.*

Note that no hypothesis of local path-connectedness is necessary here.

Proof. If $\tilde{F}$ exists, then for each $x \in X$,

$$t \to \tilde{F}(x, t)$$

is a path $I \to \mathbf{R}$ which lifts the map $t \to F(x, t)$ of I to S^1, and has $0 \to \tilde{f}_0(x)$.

Now by Theorem 6.2, these conditions do in fact determine a unique lifted path $I \to \mathbf{R}$. We will therefore denote its value at t by $\tilde{F}(x, t)$. The properties

$e \circ \tilde{F} = F$ and $\tilde{F}(x, 0) = \tilde{f}_0(x)$ follow from this definition, and we have shown uniqueness. We have still to prove that the constructed map $\tilde{F}$ is continuous.

For each $(x, t) \in X \times I$ choose $\varepsilon_{x,t}$ so that

$$d\big((x', t'), (x, t)\big) < \varepsilon_{x,t}$$

implies

$$|F(x', t') - F(x, t)| < 1.$$

The intervals $]t - \frac{1}{2}\varepsilon_{x,t}, t + \frac{1}{2}\varepsilon_{x,t}[$ form an open cover of the compact space I. Select a finite subcover, corresponding to $t_1 < t_2 < \cdots < t_n$, and let ε be the least of the $\varepsilon_i = \frac{1}{2}\varepsilon_{x,t_i}$, and $U = U_\varepsilon(x) \cap X$. We will show, by induction on i, that $\tilde{F}$ is continuous at (x, t) for $|t - t_i| < \varepsilon_i$.

We may clearly suppose that the intervals

$$]t_{i-1} - \varepsilon_{i-1}, t_{i-1} + \varepsilon_{i-1}[\qquad \text{and} \qquad]t_i - \varepsilon_i, t_i + \varepsilon_i[$$

overlap. Let u_i be a common point, and choose $u_1 = 0$. Thus we may suppose that $\tilde{F}|U \times \{u_i\}$ is continuous at (x, u_i). Now by construction,

$$F\big(U \times]t - \varepsilon_i, t + \varepsilon_i[\big)$$

is a proper subset S of S^1; write

$$e^{-1}(S) = \bigcup_{n \in \mathbf{Z}} A + n$$

as in Lemma 6.1, with notation chosen so that $\tilde{F}(x, u_i) \in A$. Since $\tilde{F}|U \times \{u_i\}$ is continuous at (x, u_i), there is a neighborhood V of x in U with $\tilde{F}\big(V \times \{u_i\}\big) \subset A$. Now for each $y \in V$,

$$\tilde{F}\big(\{y\} \times]t - \varepsilon_i, t + \varepsilon_i[\big) \subset A$$

since it is connected. Thus

$$\tilde{F}\big(V \times]t - \varepsilon_i, t + \varepsilon_i[\big) \subset A.$$

Let $e' : A \to S$ be the restriction of e; it is a homeomorphism by Lemma 6.1. Then

$$\tilde{F}|\big(V \times]t - e_i, t + \varepsilon_i[\big) = e'^{-1} \circ F|\big(V \times]t - \varepsilon_i, t + \varepsilon_i[\big),$$

a composite of two continuous maps, so is continuous. This completes the induction. ∎

Corollary *For any space X, a map $f : X \to S^1$ has a lift $\tilde{f} : X \to \mathbf{R}$ if and only if it is homotopic to a constant map.*

Necessity is trivial: since **R** is contractible, $\tilde{f}$ is nullhomotopic, hence so is $e \circ \tilde{f} = f$ (compare the proof of (iii) $\Rightarrow$ (i) of Theorem 6.4). Conversely, if $F: f_0 \simeq f$ with f_0 a constant map with image z, choose $a_0 \in \mathbf{R}$ with $e(a_0) = z$, and let $\tilde{f}_0$ be the constant map with image a_0. Then by the theorem, we can lift the homotopy F to $\tilde{F}: X \times I \to \mathbf{R}$. Then the map $\tilde{f}: X \to \mathbf{R}$ defined by $\tilde{f}(x) = \tilde{F}(x, 1)$ is a lift of f. ■

THE EXTENSION PROBLEM

We now come to the second problem mentioned at the beginning of the chapter, and associated with the diagram

$$\begin{array}{ccc} A & \xrightarrow{g} & B \\ {\scriptstyle f}\downarrow & \swarrow {\scriptstyle h} & \\ C. & & \end{array}$$

Clearly, if h exists, $g(x) = g(y)$ implies

$$f(x) = h(g(x)) = h(g(y)) = f(y).$$

A reasonable hypothesis to make is thus that g should be injective; it is more convenient to suppose g an embedding. Furthermore if, say, $a \in A$ is the limit of a sequence $\{b_n\}$ of points of B, it is necessary to define $h(a)$ as the limit of $\{f(b_n)\}$, supposing this limit to exist. To avoid such analytic problems, it is wise to assume that A is in fact a closed subspace of B. We have thus restricted consideration to the problem of whether the map f defined on the closed subspace A of B has a continuous extension over B. As we have avoided the analytical difficulties of the factorization problem, this restricted problem is primarily algebraic in nature.

In this chapter, we give an important result which solves the extension problem in a particular case.

7.4 Theorem (*Tietze's extension theorem*). *Let B be a space, A a closed subspace, J a closed interval of real numbers. Then any continuous map $f: A \to J$ has a continuous extension to a map $F: B \to J$.*

All closed intervals are homeomorphic, so we may replace J by the interval $[-1, 1]$. First we prove a lemma.

7.5 Lemma *Let $h: A \to \mathbf{R}$ have $|h(x)| \leqslant k$ for all $x \in A$. Then there is a continuous map $H: B \to \mathbf{R}$ such that $|H(x)| \leqslant \frac{1}{3}k$ for $x \in B$ and $|h(x) - H(x)| \leqslant \frac{2}{3}k$ for $x \in A$.*

Proof. Let

$$A^{+} = h^{-1}[\tfrac{1}{3}k, k] \quad \text{and} \quad A^{-} = h^{-1}[-k, -\tfrac{1}{3}k];$$

these are closed subsets of A since h is continuous. Since A is closed in B, so are A^{-} and A^{+}. By Lemma 1.4, $x \to d(x, A^{+})$ is a continuous function on B to $\mathbf{R}_{+}$, vanishing only on A^{+}; similarly for A^{-}. Thus $d(x, A^{+}) + d(x, A^{-})$ is never zero on B and the function

$$H(x) = \tfrac{1}{3}k\frac{d(x, A^{-}) - d(x, A^{+})}{d(x, A^{-}) + d(x, A^{+})}$$

is well defined and continuous on B; clearly $|H(x)| \leqslant \frac{1}{3}k$.

Now if $x \in A$ and $h(x) \geqslant \frac{1}{3}k$, then $x \in A^{+}$. Then $d(x, A^{+}) = 0$, so $H(x) = \frac{1}{3}k$ and

$$0 \leqslant h(x) - H(x) \leqslant \tfrac{2}{3}k.$$

Similarly, if $h(x) \leqslant -\frac{1}{3}k$. But if $-\frac{1}{3}k \leqslant h(x) \leqslant \frac{1}{3}k$ then

$$|H(x) - h(x) \leqslant |H(x)| + |h(x)| \leqslant \tfrac{2}{3}k. \;\blacksquare$$

Proof of Theorem 7.4. We are given $f\colon A \to [-1, 1]$. By the lemma, with $k = 1$, $h = f$, we can find $F_1 : B \to [-\frac{1}{3}, \frac{1}{3}]$, with $|f(x) - F_1(x)| \leqslant \frac{2}{3}$ for $x \in A$. We can now apply the lemma with $k = \frac{2}{3}$ and $h = f - F_1|A$. Suppose inductively that maps $F_i\colon B \to \mathbf{R}$ have been constructed with $|F_i(x)| \leqslant 2^{i-1}/3^i$ on B and

$$\Big|f(x) - \sum_{r=1}^{i} F_r(x)\Big| \leqslant (\tfrac{2}{3})^i \qquad \text{for} \quad x \in A,$$

for $i \leqslant n - 1$. Apply the lemma with

$$h = f - \sum_{r=1}^{n-1} (F_r|A) \qquad \text{and} \qquad k = (\tfrac{2}{3})^{n-1}:$$

we obtain a map F_n satisfying the corresponding conditions. By induction, we may suppose F_i constructed for all i. Then since $|F_i(x)| \leqslant 2^{i-1}/3^i$, the sequence ΣF_i converges uniformly on B (by the Weierstrass M-test), so (see below) the sum is a well-defined, continuous map $F\colon B \to \mathbf{R}$, with

$$|F(x)| \leqslant \sum_{i=1}^{\infty} \frac{2^{i-1}}{3^i} = 1.$$

Making i tend to infinity in the inequality, we see that $f(x) = F(x)$ for $x \in A$. Thus F satisfies the requirements of the theorem. $\blacksquare$

For the benefit of readers unfamiliar with the notion of uniform convergence, we give here the full proof that the map F defined in the proof of Theorem 7.4 as ΣF_i is continuous.

Given $b \in B$ and $\varepsilon > 0$, we must find $\delta > 0$ such that $b' \in B$, $d(b, b') < \delta$ implies $|F(b) - F(b')| < \varepsilon$. Choose n so that $(\frac{2}{3})^n < \varepsilon/3$. Then for all $b' \in B$,

$$|F(b') - \sum_1^n F_i(b')| \leqslant (\tfrac{2}{3})^n < \varepsilon/3.$$

Since $\sum_1^n F_i$ is continuous, there exists δ such that $d(b, b') < \delta$ implies $|\sum_1^n (F_i(b) - F_i(b')| < \varepsilon/3$. But this implies in turn that $|F(b) - F(b')| < \varepsilon$, which is what we had to prove.

Corollary *Let J be an open interval on $\mathbf{R}$; A and B as in the theorem. Then any continuous map $f: A \to J$ has a continuous extension $F: B \to J$.*

Proof. Since all open intervals (whether of finite or infinite length) are homeomorphic (see Exercise 1.5), we may suppose $J =]-1, 1[$.

By Theorem 7.4, f has an extension $g: B \to [-1, 1]$. Let $C = g^{-1}\{-1, 1\}$. Then C is a closed subspace of B, disjoint from A. Define $h: A \cup C \to I$ by $h(A) = 1$, $h(C) = 0$. This is continuous since A and C are closed. By Theorem 7.4 again, it has a continuous extension $k: B \to I$. We can now define F by setting $F(x) = g(x)k(x)$. ∎

7.6 Proposition *Let B be a space, A a closed subspace. Let $f_0: B \to S^1$, and let $g_t: A \to S^1$ be a homotopy with $g_0 = f_0|A$. Then g_t can be extended to a homotopy f_t of f_0.*

Proof. Define h_t by $h_t(a) = g_t(a)/g_0(a)$. This is a homotopy of the constant map $A \to 1$; by Theorem 7.2 it lifts to a homotopy $A \times I \to \mathbf{R}$ of the constant map $A \to 0$. By the corollary to Theorem 7.4, since $X \times 0 \cup A \times I$ is a closed subset of $X \times I$, we can extend this to a homotopy $X \times I \to \mathbf{R}$ of the constant map $X \to 0$. This projects to a homotopy k_t. Now set $f_t = k_t \cdot f_0$. ∎

Corollary *Let B be a space, A a closed subspace, and $g: A \to S^1$ a nullhomotopic map. Then g extends to a continuous map $f: B \to S^1$.*

Take f_0 and g_0 to be constant maps in Proposition 7.6. Or lift to a map $A \to \mathbf{R}$, extend to B by the corollary to Theorem 7.4, and project down again. ∎

FURTHER DEVELOPMENTS

Our discussion of the lifting problem centers on the covering homotopy property (CHP) for e, and much of it generalizes without change to general "covering spaces" (see Exercise 6.9, for example, and Spanier or Massey); some to arbitrary maps with CHP ("fibrations"). The last result, Proposition

7.6 (the HEP), plays a dual role for the extension problem. A fuller discussion of these properties and their role in lifting and extension problems can be found for example in Hu, Spanier or Steenrod notes.

Hu, S. T., *Homotopy Theory*, Academic Press, New York, 1959.

Massey, W. S., *Algebraic Topology: An Introduction*, Harcourt, Brace and World, New York, 1967.

Spanier, E. H., *Algebraic Topology*, McGraw-Hill, New York, 1966.

"Steenrod Notes" refers to some mimeographed notes by N. E. Steenrod on *Cohomology operations and obstructions to extending continuous functions*, which used to be available from Princeton University but are now, as far as I know, unobtainable. They are beautifully written.

EXERCISES AND PROBLEMS

1. Give an example of a space B, a closed subspace A, and a map $f: A \to S^1$ which does not extend to a continuous map $B \to S^1$.
2. Prove the results corresponding to Theorem 7.4, but with J replaced by
   a) the half-open interval $[0, 1[$;
   b) the closed square in $\mathbf{R}^2$, $0 \leqslant x_1 \leqslant 1, 0 \leqslant x_2 \leqslant 1$;
   c) the open square in $\mathbf{R}^2$, $0 < x_1 < 1, 0 < x_2 < 1$;
   d) $\mathbf{R}^n$;
   e) D^n.
3. The space X is the union of three straight line segments in $\mathbf{R}^2$:
   $$x = 0, \quad -2 \leqslant y \leqslant 1; \qquad y = -2, \quad 0 \leqslant x \leqslant 1; \qquad x = 1, \quad -2 \leqslant y \leqslant 1,$$
   and the curve
   $$y = \sin(x^{-1}), \qquad 0 < x \leqslant 1.$$
   Construct a map $f: X \to S^1$ such that (i) for every loop $l: S^1 \to X$, $\deg(f \circ l) = 0$, but (ii) there is no continuous lift $\tilde{f}: X \to \mathbf{R}$. Prove that your map f satisfies (i) and (ii).
4. Suppose the continuous map $f: X \to Y$ is such that for every loop $m: S^1 \to Y$, there exists a loop $l: S^1 \to X$ with $m \simeq f \circ l$. Let Y be l.p.c., and $g: Y \to S^1$ a map such that $g \circ f$ has a continuous lift $X \to \mathbf{R}$. Then g has a continuous lift. Prove this statement. What conditions do you need to show that if g has a lift, then so has $g \circ f$?
5. Reformulate the proof of Theorem 7.3 by letting $l: S^1 \to X \times Y$ have component maps λ and μ, giving a homotopy of l to
   $$(\lambda \times \{y_0\}) * (\{x_0\} \times \mu),$$
   defined as in Exercise 6.2, and then using that exercise to compute deg. l.

6. Say that the subspace $B \subset A$ has the HEP for X if for any map $f: A \to X$ and homotopy $h_t: B \to X$ of $h_0 = f|B$, there is a homotopy f_t of $f_0 = f$ with $f_t|B = h_t$.
   a) Show that if $B \subset A$ has the HEP for X, and if $f \simeq g: B \to X$, then if either f or g extends to a map $A \to X$, so does the other.
   b) Show that if $B \subset A$ has the HEP for all spaces, there exist a neighborhood U of B in A and a map $r: U \to B$ with $r|B$ the identity. [*Hint:* Take $X = A \times 0 \cup B \times I$, define f and h_t in the obvious way, and use the map f_1.]

8
CALCULATIONS

INTRODUCTION

After the hard preparatory work in the preceding chapter, we are now ready to give examples of calculations of the groups $H^1(X)$. We first prove a general result which will be our main tool. This takes the form of an exact sequence: this is typical of results in algebraic topology. The group we want is not determined explicitly, but sufficient information is given for the determination to be possible in virtually any case of interest, using one or two tricks. We illustrate some such tricks by simple examples, and conclude by computing H^1 of a product space.

THE MAYER-VIETORIS THEOREM

We fix the following notation for this section. Y is a space; X_1, X_2 are *closed* subspaces with $X_1 \cup X_2 = Y$; write $W = X_1 \cap X_2$ (which is also closed). Denote the inclusion maps of the various subspaces by

$$\begin{array}{ccccc} & & X_1 & & \\ & \overset{i_1}{\nearrow} & & \overset{j_1}{\searrow} & \\ W & & \xrightarrow{\quad k \quad} & & Y \\ & \underset{i_2}{\searrow} & & \underset{j_2}{\nearrow} & \\ & & X_2 & & \end{array}$$

so that $k = j_1 \circ i_1 = j_2 \circ i_2$.

8.1 Theorem *There is a homomorphism*

$$\delta^* : H^0(W) \to H^1(Y)$$

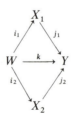

such that the following sequence is exact:

$$0 \longrightarrow H^0(Y) \xrightarrow{\{j_1^*, -j_2^*\}} H^0(X_1) \oplus H^0(X_2) \xrightarrow{(i_1^*, i_2^*)} H^0(W) \xrightarrow{\delta^*}$$

$$\xrightarrow{\delta^*} H^1(Y) \xrightarrow{\{j_1^*, -j_2^*\}} H^1(X_1) \oplus H^1(X_2) \xrightarrow{(i_1^*, i_2^*)} H^1(W).$$

Proof

Exactness at $H^0(Y)$. Let $h\colon Y \to \mathbf{Z}$ represent an element of Ker $\{j_1^*, -j_2^*\}$. Then

$$0 = j_1^*(h) = h \circ j_1 = h|X_1.$$

Similarly, $h|X_2 = 0$. Since $Y = X_1 \cup X_2$, then $h = 0$.

Exactness at $H^0(X_1) \oplus H^0(X_2)$. We have

$$(i_1^*, i_2^*)\{j_1^*, -j_2^*\} = i_1^*j_1^* - i_2^*j_2^* = k^* - k^* = 0.$$

Conversely, let (g_1, g_2) be an element of the kernel. Then

$$0 = i_1^*(g_1) + i_2^*(g_2) = g_1|W + g_2|W.$$

We define $h\colon Y \to \mathbf{Z}$ by $h|X_1 = g_1$, $h|X_2 = -g_2$. Since these agree on $X_1 \cap X_2 = W$, h is well defined. It is continuous by Theorem 1.7. And $g_1 = h|X_1 = j_1^*(h)$, $g_2 = -h|X_2 = -j_2^*(h)$, so the pair $(g_1, g_2) = \{j_1^*, -j_2^*\}(h)$.

The rest of the argument is of a similar nature, but is more involved.

Definition of δ^*. Let $f\colon W \to \mathbf{Z}$ be a continuous map. By the corollary to Theorem 7.4, we can extend to a continuous map $g\colon X_1 \to \mathbf{R}$. We define $h\colon Y \to S^1$ by

$$h|X_1 = e \circ g, \qquad h(X_2) = \{1\}.$$

These agree on W (since $g(W) = f(W) \subset \mathbf{Z}$); each is continuous, hence, by Theorem 1.7, so is h.

We must show that the homotopy class of h does not depend on the choice of the extension g. Let g_0 and g_1 be two extensions. We define a homotopy by

$$g_t(x) = (1 - t)g_0(x) + tg_1(x).$$

It is easy to check that this is continuous and that $g_t|W = f$. Then defining h_t by

$$h_t|X_1 = e \circ g_t, \qquad h_t(X_2) = \{1\}$$

gives the desired homotopy between h_0 and h_1.

We can thus define $\delta^*(f)$ unambiguously as the homotopy class of h. It is immediate that δ^* is a homomorphism.

Exactness at $H^0(W)$. To prove that $\delta^* \circ (i_1^*, i_2^*) = 0$ is equivalent to showing $\delta^* \circ i_1^* = 0$ and $\delta^* \circ i_2^* = 0$. Now if $g_1 : X_1 \to \mathbf{Z}$ restricts to $f = i_1^*(g_1): W \to \mathbf{Z}$, we can use the extension g_1 of f in computing $\delta^*(f)$. The resulting map h is then a constant map, so $0 = \delta^*(f) = \delta^* i_1^*(g_1)$. Next suppose that f is the restriction of $g_2 : X_2 \to \mathbf{Z}$. Form $g: X_1 \to \mathbf{R}$, extending f, as before. Then g and g_2 agree on W, so give a continuous map $\tilde{h}: Y \to \mathbf{R}$; moreover $e \circ \tilde{h} = h$. Since h has a continuous lift, it is nullhomotopic. This shows that $\delta^* \circ i_2^* = 0$.

Now, conversely, let $f: W \to \mathbf{Z}$ be such that the map $h: Y \to S^1$, defined as above using an extension $g: X_1 \to \mathbf{R}$ of f, is nullhomotopic. By Theorem 7.3, we can find a continuous lift $\tilde{h}: Y \to \mathbf{R}$. Since $h(X_2) = \{1\}$, $\tilde{h}(X_2) \subset \mathbf{Z}$, so $\tilde{h}$ defines by restriction a map $g_2 : X_2 \to \mathbf{Z}$. Also $\tilde{h}|X_1$ and g both lift the same map $h|X_1 = e \circ g$. Thus we can define a continuous map $g_1 : X_1 \to \mathbf{Z}$ by

$$g_1(y) = g(y) - \tilde{h}(y).$$

But for $w \in W$, g_1 and g_2 are both defined, and

$$f(w) = g(w) = g_1(w) + g_2(w).$$

Hence $f = (i_1^*, i_2^*)(g_1, g_2)$.

Exactness at $H^1(Y)$. We begin by showing that $j_1^* \circ \delta^* = j_2^* \circ \delta^* = 0$. For let f, g, h be as in the definition of δ^*. Then $j_2^*\delta^*(f)$ is represented by the constant map $h|X_2$, so is zero, whereas $j_1^*\delta^*(f)$ is represented by $h|X_1 = e \circ g$ which is nullhomotopic because it has a lift.

Now suppose given an element of $H^1(Y)$ which is annihilated by $\{j_1^*, -j_2^*\}$, that is by both j_1^* and j_2^*. It is then represented by a map $h: Y \to S^1$ whose restrictions to X_1 and X_2 are both nullhomotopic. By Proposition 7.6, the homotopy of $h|X_2$ to the constant map $X_2 \to \{1\}$ can be extended to a homotopy of h. We can thus replace h by a homotopic map $h': Y \to S^1$ with $h'(X_2) = \{1\}$. Now $h'|X_1$ is homotopic to $h|X_1$, so is again nullhomotopic. Hence it has a continuous lift $g: X_1 \to \mathbf{R}$. If $f = g|W$, then f lifts the map $W \to \{1\}$, so takes values in $\mathbf{Z}$. It is now clear that $\delta^*(f)$ is represented by h', hence equals the class of h', or of h.

Exactness at $H^1(X_1) \oplus H^1(X_2)$. The argument here is very similar to the zero-dimensional case. First we have

$$(i_1^*, i_2^*)\{j_1^*, -j_2^*\} = i_1^* j_1^* - i_2^* j_2^* = k^* - k^* = 0.$$

Next, let $g_1 : X_1 \to S^1$ and $g_2 : X_2 \to S^1$ represent an element of the kernel. Then the classes in $H^1(W)$ defined by $g_1|W$ and $g_2|W$ have sum zero, so $g_1|W$ is homotopic to $(g_2|W)^{-1}$. By Proposition 7.6 again we can extend this homotopy to one of g_1, and thus replace g_1 by a homotopic map $g'_1 : X_1 \to S^1$ with

$$g'_1(w) = (g_2(w))^{-1} \qquad \text{for all} \quad w \in W.$$

Now define $h: Y \to S^1$ by $h|X_1 = g'_1$, and

$$h(y) = (g_2(y))^{-1} \qquad \text{for} \quad y \in X_2.$$

By Theorem 1.7, this gives a continuous map, and by definition $\{j_1^*, -j_2^*\}(h)$ is the class defined by (g'_1, g_2) or equivalently, by (g_1, g_2).

We have now proved all parts of the theorem. ■

FIRST CALCULATIONS

We now give some simple examples to illustrate how this result can be used to perform computations. It is perhaps worth emphasizing that calculations in general are usually ingenious combinations of just such simple steps.

8.2 Lemma *Let $X = X_1 \cup X_2$ be a partition of X. Then there are isomorphisms*

$$H^0(X) \cong H^0(X_1) \oplus H^0(X_2),$$

$$H^1(X) \cong H^1(X_1) \oplus H^1(X_2).$$

Proof. By hypothesis, each of X_1 and X_2 is closed in X. Thus we can apply Theorem 8.1 to obtain an exact sequence. Since $X_1 \cap X_2$ is empty, $H^0(X_1 \cap X_2)$ and $H^1(X_1 \cap X_2)$ are zero, so we have exact sequences

$$0 \to H^0(X) \to H^0(X_1) \oplus H^0(X_2) \to 0,$$

$$0 \to H^1(X) \to H^1(X_1) \oplus H^1(X_2) \to 0.$$

The result follows by a simple property of exact sequences. ■

Of course, in this case it would have been easy to prove the result directly; however, the result is a useful one to have in explicit form (note that it extends by induction for splitting into a finite set of components), and the method extends to the next cases. In fact we now consider the case when $X_1 \cap X_2 = \{P\}$ consists of just one point. Since here each (continuous) map $\{P\} \to \mathbf{Z}$ extends to an (e.g. constant) map $X_1 \to \mathbf{Z}$, $H^0(X_1) \to H^0(P)$ is surjective. By exactness, the map

$$\delta^* : H^0(P) \to H^1(X_1 \cup X_2)$$

is zero. Using exactness again as above, we deduce the first part of

8.3 Lemma *Suppose X_1, X_2 are closed in X with union X and intersection a*

single point P: when this is the case, we write $X = X_1 \vee X_2$. Then there are isomorphisms

$$H^1(X) = H^1(X_1 \vee X_2) \cong H^1(X_1) \oplus H^1(X_2),$$
$$H^0(X) \oplus \mathbf{Z} \cong H^0(X_1) \oplus H^0(X_2).$$

If also X_2 is connected, then $H^0(X) \cong H^0(X_1)$.

For in the exact sequence

$$0 \to H^0(X) \to H^0(X_1) \oplus H^0(X_2) \to H^0(P) \xrightarrow{0},$$

$H^0(P)$ is isomorphic to $\mathbf{Z}$, hence free, so the sequence splits by Proposition 2.7. If X_2 is connected, the map $H^0(X_2) \to H^0(P)$ is an isomorphism. The result now follows from the corollary to Theorem 2.6, or can be shown directly. ■

More interesting is the case when $X_1 \cap X_2$ contains two points. The most important examples are covered by

8.4 Theorem *Let $Y = X \cup A$ where X is closed in Y, A is an arc with endpoints P, Q and $X \cap A = \{P, Q\}$. If X has a partition which separates P from Q, then*

$$H^0(X) \cong H^0(Y) \oplus \mathbf{Z} \qquad \textit{and} \qquad H^1(Y) \cong H^1(X).$$

Otherwise, we have an isomorphism $H^0(Y) \cong H^0(X)$ and an exact sequence

$$0 \to \mathbf{Z} \to H^1(Y) \to H^1(X) \to 0.$$

If there is a path in X joining P to Q (for example if X is l.p.c.), this sequence splits, so

$$H^1(Y) \cong H^1(X) \oplus \mathbf{Z}.$$

Proof. Since A is an arc, it is compact, hence closed by Lemma 1.9. Thus we can apply Theorem 8.1. Now $H^1(A) = H^1(P) = 0$, and by Lemma 8.2,

$$H^0(X \cap A) \cong H^0(P) \oplus H^0(Q) \cong \mathbf{Z} \oplus \mathbf{Z},$$
$$H^1(X \cap A) \cong H^1(P) \oplus H^1(Q) = 0.$$

Thus the exact sequence has the form

$$0 \to H^0(Y) \to H^0(X) \oplus H^0(A) \to H^0(P) \oplus H^0(Q) \xrightarrow{\delta^*} H^1(Y) \to H^1(X) \to 0.$$

But (since A is contractible) the restriction map $H^0(A) \to H^0(Q)$ is an isomorphism. Then we can apply Theorem 2.6 to "cancel" these terms and obtain

$$0 \to H^0(Y) \to H^0(X) \xrightarrow{\lambda} \mathbf{Z} \xrightarrow{\delta} H^1(Y) \to H^1(X) \to 0,$$

where for $f: X \to \mathbf{Z}$ we have $\lambda(f) = f(P) - f(Q)$.

We now distinguish the two cases. First suppose there is a partition $X = X_1 \cup X_2$ with $P \in X_1$, $Q \in X_2$ (Fig. 8.1). Then any map $g: \{P, Q\} \to \mathbf{Z}$ can be extended to X by defining $g(X_1) = g(P)$, $g(X_2) = g(Q)$. In particular, λ is surjective. By exactness, $\delta = 0$. The result now follows as in Lemma 8.3. Indeed, an alternative proof for this case can be obtained by applying the lemma in turn to $X_1 \vee A$ and to $(X_1 \cup A) \vee X_2$.

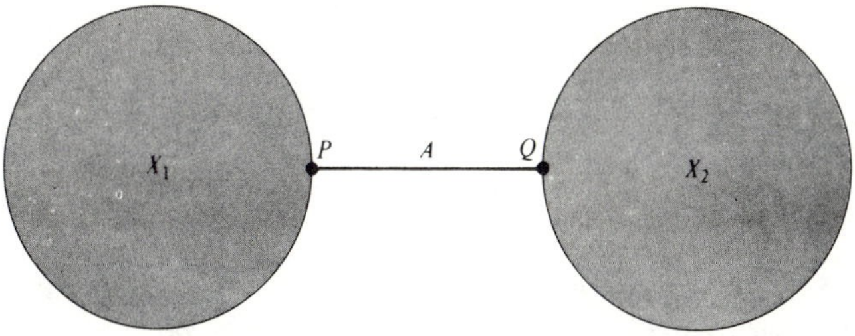

Fig. 8.1

Now suppose there is no such partition (Fig. 8.2). Then $\lambda = 0$, for if $f: X \to \mathbf{Z}$ with $f(P) \neq f(Q)$, we could obtain a partition by defining $X_1 = f^{-1}(f(P))$. The second assertion of the theorem follows.

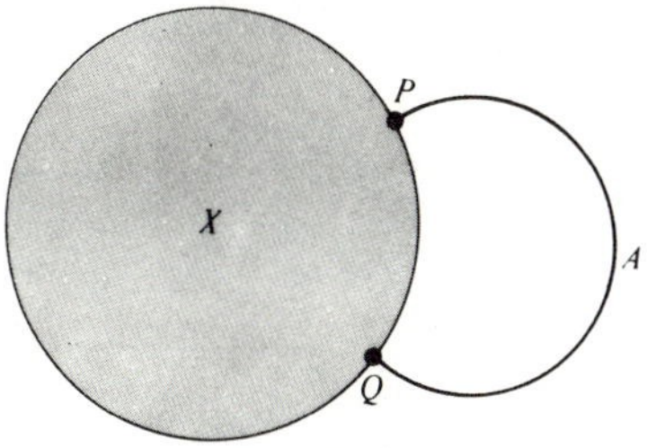

Fig. 8.2

Finally, suppose there is a path in X joining P to Q. We can regard this as an arc B with endpoints P and Q and a map $i: B \to X$. Clearly we can suppose $A \cap B = \{P, Q\}$. Then i extends to a map

$$j: A \cup B \to A \cup X = Y,$$

and since $A \cup B$ is homeomorphic to S^1, we have

$$j^*: H^1(Y) \to H^1(A \cup B) \cong H^1(S^1) \cong \mathbf{Z}.$$

I claim that this is (up to sign) left inverse to δ, so by Exercise 2.5, the sequence is split.

Indeed, applying what we proved above, but with B in place of X, gives an exact sequence:

$$0 \to \mathbf{Z} \xrightarrow{\delta'} H^1(A \cup B) \to H^1(B) \to 0.$$

Since B is an arc, $H^1(B) = 0$ and δ' is an isomorphism. Now consider

$$\begin{array}{ccccccccc} 0 & \to & \mathbf{Z} & \xrightarrow{\delta} & H^1(A \cup X) & \to & H^1(X) & \to & 0 \\ & & \| & & \downarrow{\scriptstyle j^*} & & \downarrow{\scriptstyle i^*} & & \\ 0 & \to & \mathbf{Z} & \xrightarrow{\delta'} & H^1(A \cup B) & \to & H^1(B) & \to & 0 \end{array}$$

From the definition of δ, it is easy to see that $j^* \circ \delta = \delta'$. But δ' is an isomorphism. Hence $(\delta')^{-1} \circ j^*$ is indeed left inverse to δ. ■

GRAPHS

In the situation of Theorem 8.4 above, namely $Y = X \cup A$ with A an arc, X closed in Y, and $X \cap A$ just the two endpoints of A, we say Y is obtained from X by *attaching an arc*. We now define, inductively, a (finite) graph to be a space obtained from a finite set of points $P_1, \ldots, P_N$ (the vertices) by attaching a finite number of arcs, one after the other, with the endpoints of the arcs among the points P_i (Fig. 8.3).

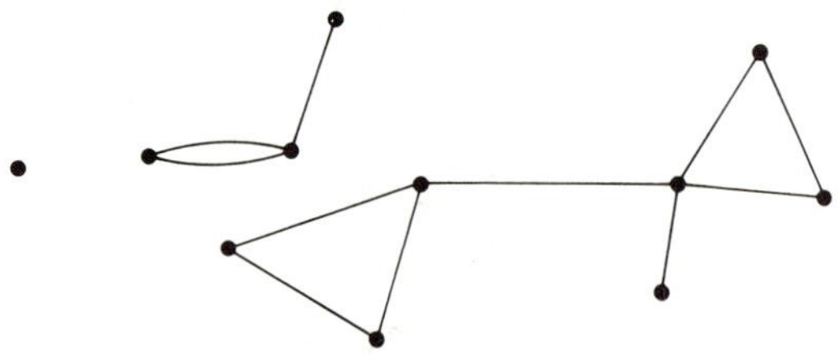

Fig. 8.3

Certain properties of graphs G follow from the inductive construction. First, G is compact (Exercise 1.14). Next, G is l.p.c. (Exercise 3.12). Then G has a finite number of path components (clear, since an arc is path-connected).

Finally, it follows inductively from Theorem 8.4 that $H^1(G)$ is free abelian of finite rank. We will show how to calculate this rank which, by Proposition 2.4, is well determined.

First consider $\pi_0(G)$. Since arcs are path-connected, each path component contains at least one of the vertices P_i. Next, two vertices P_i, P_j are in the same path component if they are joined by an arc P_iP_j, or more generally if there is a chain of arcs such as

$$P_iP_r, \quad P_rP_s, \quad P_sP_j$$

joining them. Conversely, if we take the union of the vertices which can be so joined to P_i, and of the arcs through them, we have an open set in G, hence a path component. So the determination of $\pi_0(G)$ is straightforward.

8.5 Theorem *Let G be a graph with α_0 vertices joined by α_1 arcs. Let G have β_0 components, and let β_1 be the rank of the free abelian group $H^1(G)$. Then*

$$\alpha_0 - \alpha_1 = \beta_0 - \beta_1.$$

Thus to determine β_1 we need only find $\pi_0(G)$ as above and then count vertices and edges.

Proof. By induction on α_1. If $\alpha_1 = 0$, G consists simply of the α_0 points P_i. Since $H^1(P_i) = 0$, it follows by Lemma 8.2 that $\beta_0 = \alpha_0$ and $\beta_1 = 0 = \alpha_1$.

Now suppose G obtained from G' by attaching an arc; it will be enough to assume the result for G' and deduce it for G. Denote β_0 for G' by β_0'; similarly for the others. Then

$$\alpha_0 = \alpha_0', \qquad \alpha_1 = \alpha_1' + 1$$

by inductive construction, and

$$\alpha_0' - \alpha_1' = \beta_0' - \beta_1'$$

by the inductive hypothesis. Now by Theorem 8.4,

$$\begin{aligned} \text{either} \quad & \beta_0 = \beta_0' - 1, \quad \beta_1 = \beta_1' \\ \text{or} \quad & \beta_0 = \beta_0', \qquad \beta_1 = \beta_1' + 1. \end{aligned}$$

It follows at once that in either case we have

$$\alpha_0 - \alpha_1 = \beta_0 - \beta_1. \blacksquare$$

PRODUCTS

There is a natural bijection between $\pi_0(X \times Y)$ and $\pi_0(X) \times \pi_0(Y)$ for any two spaces X and Y (Exercise 4.3). If also X and Y are l.p.c., then so (Exercise

3.12) is $X \times Y$, and we can deduce the form of $H^0(X \times Y)$. For example, suppose $\pi_0(X)$ and $\pi_0(Y)$ finite (this holds, by Exercise 3.11, if X and Y are compact), having r and s elements respectively. Then $\pi_0(X \times Y)$ has rs elements, and $H^0(X)$, $H^0(Y)$, and $H^0(X \times Y)$ are free abelian groups having the respective ranks r, s, and rs.

For H^1 we begin by supposing, in addition to X and Y being l.p.c., that both are connected. Choose points $x_0 \in X$, $y_0 \in Y$ and define injections

$$i_1 : X \to X \times Y, \qquad i_2 : Y \to X \times Y$$

by

$$i_1(x) = (x, y_0), \qquad i_2(y) = (x_0, y).$$

Then we have induced maps which we can fit together as

$$(i_1^*, i_2^*) : H^1(X \times Y) \to H^1(X) \oplus H^1(Y).$$

8.6 Theorem *If X, Y are connected and l.p.c., the map*

$$(i_1^*, i_2^*) : H^1(X \times Y) \to H^1(X) \oplus H^1(Y)$$

is an isomorphism.

Proof. In the notation just introduced, Theorem 7.2 states that this map is injective. But surjectivity is easy. For given maps $f : X \to S^1$, $g : Y \to S^1$ if we define $F : X \times Y \to S^1$ by

$$F(x, y) = f(x)g(y),$$

then clearly $i_1 \circ F \simeq f$ and $i_2 \circ F \simeq g$. In fact, if we had required $f(x_0) = g(y_0) = 1$, we would have equality here. ■

In the more general case where X and Y are l.p.c. but have finite numbers of path components, we still obtain a result. For if we enumerate the components as

$$X_1, \ldots, X_r, \qquad Y_1, \ldots, Y_s,$$

then $X \times Y$ is l.p.c. with path components $X_i \times Y_j$ $(1 \leqslant i \leqslant r,\ 1 \leqslant j \leqslant s)$. We now have, by Lemma 8.2,

$$H^1(X \times Y) = \text{direct sum of } H^1(X_i \times Y_j), \qquad 1 \leqslant i \leqslant r, \quad 1 \leqslant j \leqslant s.$$

But by Theorem 8.6,

$$H^1(X_i \times Y_j) \cong H^1(X_i) \oplus H^1(Y_j).$$

So $H^1(X \times Y)$ is a big direct sum, in which each $H^1(X_i)$ is counted s times and each $H^1(Y_j)$ is counted r times. Recalling that (by Lemma 8.2 again) $H^1(X)$ is the direct sum of the $H^1(X_i)$, we conclude

Corollary *If X, Y are l.p.c. and have r, s path components, then $H^1(X \times Y)$ is the direct sum of s copies of $H^1(X)$ and r copies of $H^1(Y)$.* ■

Finally, write $r = \beta_0(X)$ and suppose that $H^1(X)$ is free abelian of rank $\beta_1(X)$, and correspondingly for Y. Then $H^1(X \times Y)$ is free abelian, and with this notation,

$$\beta_0(X \times Y) = \beta_0(X)\beta_0(Y),$$

$$\beta_1(X \times Y) = \beta_0(X)\beta_1(Y) + \beta_1(X)\beta_0(Y).$$

FURTHER DEVELOPMENTS

All the methods and results in this chapter have natural extensions to higher dimensions; see any book on algebraic topology, e.g. Spanier or Maunder. The analogue of the final result on products is

$$\beta_n(X \times Y) = \sum_{i+j=n} \beta_i(X)\beta_j(Y).$$

Maunder, C. R. F., *Algebraic Topology*, Van Nostrand, Princeton, 1970.

Spanier, E. H., *Algebraic Topology*, McGraw-Hill, New York, 1966.

EXERCISES AND PROBLEMS

1. Let $Y = S^1$, and let X_1, X_2 be the semicircles $\text{Im } z \geqslant 0$ and $\text{Im } z \leqslant 0$. Calculate all groups and homomorphisms in the sequence Theorem 8.1, and verify that the sequence is exact.
2. Suppose in Theorem 8.1 that X_2 and Y are connected. Show that

$$0 \to H^0(X_1) \to H^0(W) \xrightarrow{\delta^*} H^1(Y)$$

   is exact.
3. a) Let X_1 and X_2 be contractible closed subsets of a euclidean space. Show that $H^1(X_1 \cup X_2) \oplus \mathbf{Z}$ is isomorphic to $H^0(X_1 \cap X_2)$.
   b) Let $X \subset \mathbf{R}^n$; let X_1 denote the union of the straight line segments joining points of $X \times 0$ to 0×1 in $\mathbf{R}^n \times \mathbf{R} = \mathbf{R}^{n+1}$; similarly X_2 with 0×-1. Show that X_1, X_2 are contractible, so (a) applies. We write $SX = X_1 \cup X_2$.

⋆4. Give an example of a space Y such that $H^1(Y)$ is not a free abelian group. (Use the preceding exercise and take $X = \mathbf{Z}$.)

5. Suppose in the definition of δ^* (Theorem 8.1) the roles of X_1, X_2 are interchanged. Show that you obtain $-\delta^*$.
6. Show that exactness of Theorem 8.1. at $H^0(Y)$ and $H^0(X_1) \oplus H^0(X_2)$ continues to hold if X_1, X_2 are open subsets of Y, and give an example where this fails with X_1 closed, X_2 open.

★7. Show that if X_1, X_2 are both open or both closed in $X = X_1 \cup X_2$, there is an exact sequence

$$H_0(X_1 \cap X_2) \to H_0(X_1) \oplus H_0(X_2) \to H_0(X) \to 0.$$

8. Prove that $H^1(S^n) = 0$ for $n > 1$. [Apply Theorem 8.1 to two hemispheres.]

9. Let $Y = X \cup A$. Suppose there is a homeomorphism $\phi: D^n \to A$ such that $\phi(S^{n-1}) = A \cap X$. Show that if $n > 2$, $H^1(Y) \cong H^1(X)$, and describe how to determine $H^1(Y)$ when $n = 2$.

10. Let $Y = X \cup A$. Suppose there is a continuous map $\phi: D^n \to A$ such that $\phi^{-1}(X) = S^{n-1}$ and $\phi|(D^n - S^{n-1})$ is injective. Obtain Y' from Y by deleting an open disk of radius $\frac{1}{2}$ concentric with D^n. Show that Y' is homotopy equivalent to X, and hence that the results of the preceding exercise are applicable.

11. M_g is obtained from a regular $4g$-gon by twisting it so as to glue its sides in pairs as indicated in Fig. 8.4(a). Using the preceding exercise or otherwise, compute $H^1(M_g)$ and $H^1(M_g - \text{point})$.

12. As the preceding exercise, but using the surface N_h obtained from a regular $2h$-gon as in Fig. 8.4(b).

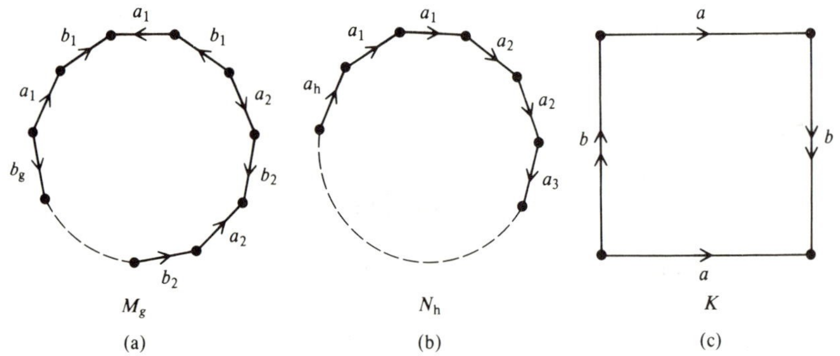

Fig. 8.4

13. As the preceding exercise, but with K; see Fig. 8.4(c). Given that K is homeomorphic to one of the M_g, N_h, determine which. (Similarly, try some further such diagrams.)

14. Let $P_j \in \mathbf{C}$ be the point $\exp(2\pi ij/n)$, $0 \leqslant j < n$. Choose ε with $0 < \varepsilon < \frac{1}{2}d(P_0, P_1)$, and let

$$X = \{z \in \mathbf{C}: |z| \leqslant 2, |z - P_i| \geqslant \varepsilon \text{ for each } i\}.$$

Let G be the graph, which is the union of the circles $|z - P_i| = \varepsilon$ and of the straight line segments joining O to $(1 - \varepsilon)P_i$. Show that the inclusion $G \subset X$ is a homotopy equivalence, and hence calculate $H^1(X)$.

15. Let G be the union of S^1 and of the straight line segments joining $\frac{1}{2}$ to the three cube roots of 1. Define $f: G \to S^1$ by projection from O. Calculate $H^1(G)$, and using this calculation, describe the map $f^*: H^1(S^1) \to H^1(G)$.

★16. With the hypothesis of the corollary to Theorem 8.6, obtain natural isomorphisms

$$H^0(X \times Y) \to H^0(X) \otimes H^0(Y),$$

$$H^1(X \times Y) \to (H^0(X) \otimes H^1(Y)) \oplus (H^1(X) \otimes H^0(Y)).$$

17. Let G be a connected finite graph. Show that the following conditions on G are equivalent:
    i) The number of vertices of G is one greater than the number of edges.
    ii) G can be built up inductively from a point by repeatedly adding an edge with one vertex in common with the part already constructed.
    iii) G is contractible.
    iv) $H^1(G)$ has only element.

    If G satisfies these conditions, G is called a *tree*.

18. A *circuit* in a graph is a finite sequence of distinct vertices, $P_0, P_1, P_2, \ldots, P_n$, such that for each i, $0 \leqslant i < n$, there is an edge P_iP_{i+1}, and there is also an edge P_nP_0. Show that a connected graph is a tree if and only if it contains no circuits.

19. G is a connected graph, T_1 a tree which is a subgraph. Show that there exists a subgraph, T, $T_1 \subseteq T \subseteq G$, such that (i) T is a tree, and (ii) T contains all the vertices of G.

20. G is a connected graph, $T \subseteq G$ a tree which contains all the vertices. Denote the other edges of G by $a_1, \ldots, a_n$; let the endpoints of a_i be P_i and Q_i. The graph Γ is the union of n triangles AB_iC_i with a common vertex A. We define $f: G \to \Gamma$ by $f(T) = A$, and $f|a_i$ is obtained by trisecting a_i at R_i and S_i, and mapping the segments P_iR_i, R_iS_i, S_iQ_i linearly onto AB_i, B_iC_i, and C_iA. We define $g: \Gamma \to G$ by mapping B_iC_i linearly onto P_iQ_i, and mapping AB_i, AC_i to paths in T (possible since T is connected). Show that $f \circ g$ is homotopic to the identity map of Γ, and that $g \circ f$ is homotopic to the identity map of G.

21. Deduce from the preceding exercise that the connected graphs G and H are homotopy equivalent if and only if $\beta_1(G) = \beta_1(H)$. Is it true that any two (finite) graphs G and H with $\beta_0(G) = \beta_0(H)$, $\beta_1(G) = \beta_1(H)$ are homotopy equivalent?

22. Construct two finite connected graphs G and H such that

$$\alpha_0(G) = \alpha_0(H), \qquad \alpha_1(G) = \alpha_1(H),$$

but G and H are not homeomorphic.

23. Show that if G and H are homeomorphic, one can subdivide by counting extra points as vertices so that there is a homeomorphism which takes vertices to vertices and edges to edges.

24. Show that by suitable choice of vertices, any finite union of straight line segments in $\mathbf{R}^n$ can be regarded as a graph.

25. Let G be a graph in $\mathbf{R}^3$ regarded as $\mathbf{R}^3 \times 0 \subset \mathbf{R}^4$, A the point $(0, 0, 0, 1)$, and C the union of the segments joining A to the vertices of G. Show that $H^1(G \cup C)$ is a free abelian group of rank $\alpha_1(G)$. Using Theorem 8.1, construct an exact sequence

$$0 \to H^0(G) \to C^0 \xrightarrow{d} C^1 \to H^1(G) \to 0,$$

where C^0, C^1 are free abelian with ranks α_0, α_1 respectively. Using Exercise 2.16, deduce Theorem 8.5.

26. In the preceding exercise, for each (directed) edge PQ of G, define $r(PQ): H^1(G \cup C) \to \mathbf{Z}$ by taking the degree of the restriction of $f: G \cup C \to S^1$ to the circle $APQA$. Show that by choosing a direction for each edge, we obtain an isomorphism of $H^1(G \cup C)$ with the free abelian group on the edges of G. Use this to compute the map $d: C^0 \to C^1$ above.

27. Obtain an alternative proof of this result as follows. Break each edge PQ as the union of three abutting arcs $PP'Q'Q$. Let X_1 denote the union of the arcs ending at a vertex of G; X_2 the union of the "middle third" arcs. Write $X_1 \cap X_2$ as the union of the sets I of initial points (for example, P') and F of final points (for example, Q'). Then Theorem 8.1 leads to an exact sequence

$$0 \to H^0(G) \to H^0(X_1) \oplus H^0(X_2) \to H^0(I) \oplus H^0(F) \to H^1(G) \to 0,$$

in which the map $H^0(X_2) \to H^0(F)$ is an isomorphism. Cancel this by Theorem 2.6 and compute the resulting map $H^0(X_1) \to H^0(I)$.

PART 2
THE DUALITY THEOREM

9
EILENBERG'S SEPARATION CRITERION

INTRODUCTION

The main aim of Part II is the application of the results of Part I to obtain more detailed information in the case of subsets of the plane. A general problem is the relation between the topology of a subset and that of its complement, and our main result shows how the number of components of the complement of a compact subset is determined by intrinsic properties of that subset. This contains as a special case the celebrated Jordan curve theorem: that if a subset of a plane is homeomorphic to S^1, its complement has two components. In this chapter we prepare for this theorem with some simple results of a general nature, and then obtain a necessary and sufficient condition (the criterion of the title) for two points of the plane to be separated by a compact subset.

It is convenient to identify the euclidean space $\mathbf{R}^2$ of points (x, y) with the space $\mathbf{C}$ of complex numbers $z = x + iy$. With this notation, the distance between the points z_1 and z_2 is just $|z_1 - z_2|$. We make frequent use of the modulus and amplitude of complex numbers, so introduce the notation

$$N(z) = z/|z|, \qquad z \neq 0.$$

Then N is a continuous map

$$N: \mathbf{C} - \{0\} \to S^1.$$

COMPLEMENTARY COMPONENTS

Our first results are valid in any euclidean space.

9.1 Lemma *If K is a compact subset of $\mathbf{R}^n$, $\mathbf{R}^n - K$ is l.p.c.*

Proof. By Lemma 1.9, K is closed in $\mathbf{R}^n$, so $\mathbf{R}^n - K$ is open. But by Lemma 3.5(i), $\mathbf{R}^n$ is l.p.c., and by Lemma 3.5(ii), so is any open subset. ■

This result shows that we do not need to worry about the different definitions of connection for $\mathbf{R}^n - K$. We refer to the path components of $\mathbf{R}^n - K$ simply as components (for the definition of components in general, see Exercise 4.8). Two points of $\mathbf{R}^n - K$ are said to be *separated* by K if they lie in different components of $\mathbf{R}^n - K$, that is, if they are not joinable by a path in $\mathbf{R}^n - K$.

9.2 Lemma *Let K be a compact subset of $\mathbf{R}^n$, A the union of some of the components of $\mathbf{R}^n - K$. Then $A \cup K$ is closed in $\mathbf{R}^n$.*

Proof. Let $B = \mathbf{R}^n - K - A$ be the union of the remaining components of $\mathbf{R}^n - K$. Since $\mathbf{R}^n - K$ is l.p.c., these are open in $\mathbf{R}^n - K$ by Lemma 4.6, hence so by Proposition 1.3 is their union B. As $\mathbf{R}^n - K$ is open in $\mathbf{R}^n$, B is also, and its complement $A \cup K$ is closed. ■

9.3 Lemma *Let K be a compact subset of $\mathbf{R}^n$. Then $\mathbf{R}^n - K$ has just one unbounded component, U say, and $\mathbf{R}^n - U$ is bounded.*

Proof. Since K is compact, by Theorem 1.8 it is bounded, say $|z| \leqslant R$ for $z \in K$. Define

$$E = \{P \in \mathbf{R}^n : d(P, 0) > R\}.$$

Clearly [Exercise 3.1(a, iii)], E is path-connected; since it is disjoint from K, it lies in some component U of $\mathbf{R}^n - K$. Then K itself and the other components of $\mathbf{R}^n - K$ do not meet E, and so are contained in the closed ball with centre O and radius R. ■

SEPARATION OF POINTS BY COMPACT PLANE SETS

9.4 Theorem (*Eilenberg's criterion*). *Let K be a compact subset of $\mathbf{C}$; a, b points of $\mathbf{C} - K$. Then a and b are in the same component of $\mathbf{C} - K$ if and only if the map $f: K \to S^1$ defined by*

$$f(z) = N\left(\frac{z - a}{z - b}\right) \quad \textit{for} \quad z \in K$$

is nullhomotopic.

Proof. First suppose that the two points are in the same component. Let $p: I \to \mathbf{C} - K$ be a path with $p(0) = a$, $p(1) = b$. Then a homotopy of f to a constant map is given by

$$H(z, t) = N\left(\frac{z - p(t)}{z - b}\right).$$

Now suppose that the two points are in different components. We will assume that f is nullhomotopic, and derive a contradiction. Let A be the component of $\mathbf{C} - K$ containing a. It will be convenient to suppose A bounded. If it is not, we can interchange a and b to obtain this boundedness (by Lemma 9.3), and f is only replaced by the inverse map $1/f$.

Since f is nullhomotopic and K closed in $\mathbf{C}$, hence in $\mathbf{C} - A$, we can apply the corollary to Proposition 7.6 to deduce that f can be extended to a continuous map

$$g_1 : \mathbf{C} - A \to S^1,$$

and that we can also assume g_1 nullhomotopic. Now the formula defining f makes sense for all $z \neq a, b$. We can thus regard it as defining a continuous extension of f:

$$g_2 : A \cup K - \{a\} \to S^1 .$$

Now g_1 and g_2 agree on the intersection K of their domains, thus define a map

$$F : \mathbf{C} - \{a\} \to S^1$$

extending f. Moreover F is continuous by Theorem 1.7 since $\mathbf{C} - A$ is closed in $\mathbf{C}$ (A is open by Lemma 9.1) and $A \cup K$ is closed in $\mathbf{C}$ by Lemma 9.2.

We can thus define a homotopy

$$H : S^1 \times I \to S^1$$

by

$$h_t(z) = H(z, t) = F\big(a + z(\varepsilon + tR)\big) \qquad |z| = 1, \quad 0 \leqslant t \leqslant 1;$$

thus by Theorem 6.4, $\deg h_0 = \deg h_1$. Now since A is bounded, for R sufficiently large, $a + z(\varepsilon + R)$ will not be in A for any $z \in S^1$. Since g_1 is nullhomotopic, so is then h_1, and $\deg h_1 = 0$.

On the other hand, since A is open, if ε is small enough, we will have $(a + z\varepsilon) \in A$ for all $z \in S^1$. Thus

$$\begin{aligned} h_0(z) &= F(a + z\varepsilon) = g_2(a + z\varepsilon) \\ &= N\left(\frac{z\varepsilon}{a + z\varepsilon - b}\right) = N(z)N(\varepsilon)N(a + z\varepsilon - b)^{-1}. \end{aligned}$$

This expresses h_0 as the product of three maps, so $\deg h_0$ is the sum of their degrees. The first, the identity, has degree 1. The second is constant, and so has degree 0. The third, for ε small, takes values only in a neighborhood of $N(a - b)^{-1}$, so is nullhomotopic by the corollary to Lemma 6.1. Adding up, we find

$$\deg h_0 = 1 \neq 0 = \deg h_1,$$

a contradiction. This proves the theorem. ■

Corollary *Let K be a compact subset of* **C**, $H^1(K) = 0$. *Then* $\mathbf{C} - K$ *is connected. This is true, in particular, for K homeomorphic to I or to* I^2.

This follows immediately from the theorem. ■

9.5 Theorem *Let A, B be compact subsets of* **C**; *a, b points of* $\mathbf{C} - A - B$. *If neither A nor B separates a from b, and if* $A \cap B$ *is connected (or empty), then* $A \cup B$ *does not separate a from b.*

Proof. Take $K = A \cup B$ in Theorem 9.4. Then f defines an element of $H^1(K)$, and if this element is zero, Theorem 9.4 shows that $A \cup B$ does not separate a from b. Also, since neither A nor B separates a from b, the restrictions of f to A and B are nullhomotopic.

Now consider the Mayer-Vietoris sequence (Theorem 8.1):

$$H^0(A \cap B) \xrightarrow{\delta^*} H^1(A \cup B) \to H^1(A) \oplus H^1(B).$$

Our class in $H^1(A \cup B)$ is in the kernel of the second of these homomorphisms, hence in the image of the first. But if $A \cap B$ is empty, $H^0(A \cap B)$ is zero; if $A \cap B$ is connected, then as in the proof of Lemma 8.3 we see that $\delta^* = 0$. In either case, our class in $H^1(A \cup B)$ is zero, and the theorem follows. ■

We will use Theorem 9.5 in Chapter 13 in one of our alternative proofs of the Jordan curve theorem. If $A \cap B$ is allowed to have two components, the result can break down: for example, if A and B are the upper and lower semicircles of S^1. For a valid generalization, however, see Exercise 14.12.

FURTHER DEVELOPMENTS

The above result will be extended in one way in the next chapter. It can also be generalized to higher dimensions: a compact subset K of $\mathbf{R}^n$ fails to separate two points of its complement if and only if an appropriately defined map $K \to S^{n-1}$ is nullhomotopic. See, for example, Hurewicz and Wallman.

Hurewicz, Witold and Henry Wallman, *Dimension Theory*, Princeton University Press, 1941.

EXERCISES AND PROBLEMS

1. Show that Lemmas 9.1 and 9.2 are true for any closed subsets K of $\mathbf{C}$.
2. Give examples to show that Lemma 9.3 and Theorem 9.4 can be false if K is closed but not bounded, and that Theorem 9.4 can be false with K bounded but not closed.
3. Does Eilenberg's criterion imply that if K is homeomorphic to S^1, it separates $\mathbf{C}$?
4. What conclusions (if any) can be drawn by applying Theorem 9.4 to the following subspaces of $\mathbf{C}$?
   a) $[0, 1[$ b) S^1 c) the y-axis
   d) $x = 0, y \geqslant 0$ e) the space Y of Theorem 3.6
   f) the space X of Exercise 7.3
5. Let S be a closed and l.p.c. subset of $\mathbf{R}^n$. Show that Lemma 9.2 holds for compact subsets K of S, and hence that the union of K and the bounded components of $S - K$ is compact.
6. Let $K = K_0 \cup PQ$, where K_0 is compact and l.p.c. and the straight line segment PQ meets K_0 only at its endpoints. Let AB be a straight line segment in $\mathbf{C} - K_0$ which crosses PQ. Show that K separates A from B if and only if P and Q can be joined by a path in K_0. [*Hint:* Use Theorems 8.4 and 9.4.]

10
THE DUALITY MAP

INTRODUCTION

Following the usual procedure in algebraic topology, we will now extend Eilenberg's criterion by feeding in some algebra to assist the geometry. Essentially the same geometrical idea then gives a stronger result, and leads to conjecture a further one, which will be established in Chapter 11.

The criterion (Theorem 9.4) relates the following two ideas:

Path connection in $\mathbf{C} - K$. We have (in Chapter 4) built from this a set $\pi_0(\mathbf{C} - K)$ and a group $H_0(\mathbf{C} - K)$.

Homotopy of maps $K \to S^1$. We used this (in Chapter 5) to define a group $H^1(K)$. Thus we may hope ultimately for some relationship between $H_0(\mathbf{C} - K)$ and $H^1(K)$. It will turn out that, with one detail amended, they become isomorphic. In fact, we can restate the conclusion of Theorem 9.4 as

$$p(a) = p(b) \qquad \text{in} \quad \pi_0(\mathbf{C} - K),$$

or equivalently,

$$i(p(a)) - i(p(b)) = 0 \text{ in } H_0(\mathbf{C} - K),$$

if and only if

$$z \to N\left(\frac{z - a}{z - b}\right) \qquad \text{defines} \qquad 0 \qquad \text{in} \qquad H^1(K).$$

Keeping this in mind, we now construct, by stages, a map which we will call the duality map.

CONSTRUCTION OF THE DUALITY MAP

We begin by defining

$$D_0 : \mathbf{C} - K \to \text{Map}\,(K, S^1)$$

by

$$D_0(a)(z) = N(z - a), \qquad z \in K, \quad a \in \mathbf{C} - K.$$

10.1 Lemma *If $a \sim b$ in $\mathbf{C} - K$,*

$$D_0(a) \simeq D_0(b): K \to S^1.$$

Proof. This is essentially the same as the first part of the proof of Theorem 9.4. If $p: I \to \mathbf{C} - K$ is a path joining a to b, then $D_0(p(t))$ is a homotopy from $D_0(a)$ to $D_0(b)$. ■

Corollary *The composite $q \circ D_0$ factors through p to give a commutative diagram:*

$$\begin{array}{ccc} \mathbf{C} - K & \xrightarrow{D_0} & \text{Map}\,(K, S^1) \\ \big\downarrow{\scriptstyle p} & & \big\downarrow{\scriptstyle q} \\ \pi_0(\mathbf{C} - K) & \xrightarrow{D_1} & H^1(K). \end{array}$$

Indeed, the lemma states exactly that the homotopy class $qD_0(a)$ depends only on the component of $\mathbf{C} - K$ in which a lies. ■

Next we have defined $H_0(\mathbf{C} - K) = F(\pi_0(\mathbf{C} - K))$. Then by the universal property, there is a unique homomorphism

$$D_2: H_0(\mathbf{C} - K) \to H^1(K)$$

with $D_1 = D_2 \circ i$. The homomorphism D_2 is the one in which we are interested.

The procedure of passing first to path components and homotopy classes and then to free abelian groups can be reversed. Applying the universal property to D_0, we obtain a homomorphism

$$D_3: F(\mathbf{C} - K) \to \text{Map}\,(K, S^1)$$

with $D_3 \circ i = D_0$. Explicitly, noting that we have additive notation for $F(\mathbf{C} - K)$ and multiplicative notation for Map (K, S^1),

$$\begin{aligned} D_3\Big(\sum_r n_r i(a_r)\Big)(z) &= \prod_r \big(N(z - a_r)\big)^{n_r} \\ &= N\Big(\prod_r (z - a_r)^{n_r}\Big). \end{aligned}$$

Thus, in particular,

$$D_3\big(i(a) - i(b)\big)(z) = N\left(\frac{z - a}{z - b}\right).$$

This shows how the above all grows out of the definition of f in Theorem 9.4.

We can summarize the relations between these various definitions in a large commutative diagram. To avoid confusion, we now denote the canonical maps by

$$i:(\mathbf{C} - K) \to F(\mathbf{C} - K), \qquad i': \pi_0(\mathbf{C} - K) \to H_0(\mathbf{C} - K).$$

10.2 Lemma *The following diagram commutes.*

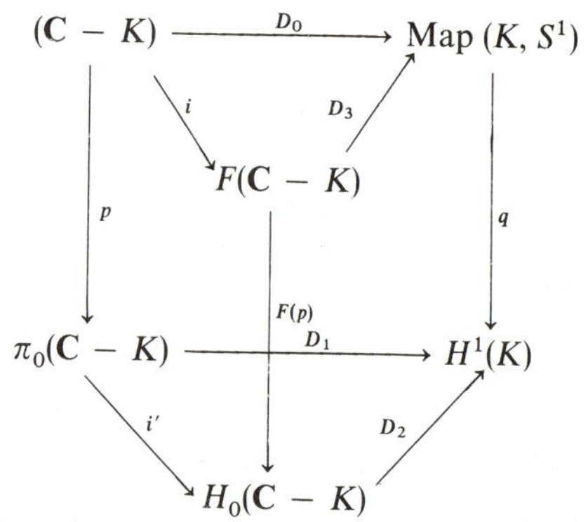

Proof. The diagram is to be pictured as a triangular prism standing on its base, and we will refer to it in geometrical terms. We have to show that each of the five faces represents a commutative diagram.

The upper triangle commutes by the definition of D_3; the lower one by the definition of D_2. The back face commutes by the definition of D_1; the front left face by the definition of $F(p)$ in Proposition 2.8. It remains to consider the front right face. But by the remarks already made,

$$q \circ D_3 \circ i = q \circ D_0 = D_1 \circ p = D_2 \circ i' \circ p = D_2 \circ F(p) \circ i;$$

and since $q \circ D_3$ and $D_2 \circ F(p)$ are homomorphisms, the universal property of i now implies that they are equal. ∎

PROOF OF INJECTIVITY

In this notation, Theorem 9.4 states: Let $a, b \in \mathbf{C} - K$ define

$$x = i'(p(a)) - i'(p(b)) \in H_0(\mathbf{C} - K).$$

Then if $D_2(x) = 0$, we have $x = 0$. This is a partial result in the direction of showing that D_2 is injective. However, D_2 is not quite injective.

10.3 Lemma *Let $K \subset \mathbf{C}$ be compact, U the unbounded component of $\mathbf{C} - K$. Then $D_1(U) = 0$.*

Proof. Suppose (as in Lemma 9.3) that $|z| \leqslant R$ for all $z \in K$. Then $2R \in U$, and for all $z \in K$,

$$D_0(2R)(z) = N(z - 2R)$$

has negative real part. By the corollary to Lemma 6.1, $D_0(2R)$ is nullhomotopic, that is, $D_1(U) = 0$. ■

We see that $H_0(\mathbf{C} - K)$ is a little too large for our needs. We cut it down to size in the following way. Let X be any nonempty space, $\{0\}$ a space with just one point, and $t: X \to \{0\}$ the unique map. Now $\pi_0(\{0\})$ has just one element, so there is an isomorphism $u: H_0(\{0\}) \to \mathbf{Z}$. Since t has a right inverse, so has $t_*: H_0(X) \to H_0(\{0\})$, which is thus surjective; in fact, for any $x \in X$,

$$u(t_*(i'(p(x)))) = 1.$$

We define $\tilde{H}_0(X)$ to be the kernel of t_*. Thus given finite sets $\{A_j\}$ of path components of X and $\{n_j\}$ of integers, $\sum n_j i(A_j) \in H_0(X)$ belongs to $\tilde{H}_0(X)$ if and only if $\sum n_j = 0$.

We denote the restriction of D_2 by

$$D: \tilde{H}_0(\mathbf{C} - K) \to H^1(K).$$

The Alexander duality theorem in the plane states that D is an isomorphism. We prove half of this now, and the rest in Chapter 11.

10.4 Theorem *For any compact subset K of $\mathbf{C}$, D is injective.*

Proof. As already promised, the proof is based on that of Theorem 9.4. Let

$$y = \sum_{j=1}^{r} n_j i'(A_j) \in H_0(\mathbf{C} - K)$$

(where all the A_j are distinct) satisfy $D(y) = 0$. We will show that $A_j \neq U$ implies $n_j = 0$, so that at most one n_j is nonzero. But if $y \in \tilde{H}_0(\mathbf{C} - K)$, we have $\sum n_j = 0$, so that all the n_j vanish, and $y = 0$.

Choose $a_j \in A_j$ for $1 \leqslant j \leqslant r$. Then $y = F(p)(x)$, with

$$x = \sum_{j=1}^{r} n_j i(a_j) \in F(\mathbf{C} - K).$$

As

$$0 = D(y) = D(F(p)(x)) = q(D_3(x)),$$

$D_3(x)$ is nullhomotopic. Hence we can apply the corollary to Proposition 7.6 to extend $D_3(x)$ to a continuous map $g_1: \mathbf{C} - A_j \to S^1$, which may be supposed nullhomotopic. Exactly as in the proof of Theorem 9.4, if we use $D_3(x)$ to define a map

$$g_2: K \cup A_j - \{a_j\} \to S^1,$$

we can fit them together to obtain a continuous map $F: \mathbf{C} - \{a_j\} \to S^1$. Again, as before, we define

$$h_t(z) = F(a_j + z(\varepsilon + tR)),$$

and note that, since A_j is bounded, provided R is large enough, h_1 is nullhomotopic, so $\deg h_0 = \deg h_1 = 0$; and further that if ε is small enough, then for $z \in S^1$,

$$h_0(z) = D_3(x)(a_j + z\varepsilon).$$

Now since $x = \sum n_k i(a_k)$ and D_3 and deg are homomorphisms, it follows that

$$0 = \deg h_0 = \sum n_k d_k,$$

where d_k is the degree of the map $f_k: S^1 \to S^1$ with

$$f_k(z) = N(a_j + z\varepsilon - a_k).$$

But as for Theorem 9.4, we see that $d_k = 0$ for $k \neq j$, while if $k = j$,

$$f_j(z) = N(z\varepsilon) = N(z)N(\varepsilon),$$

so $d_j = 1$. Hence, finally,

$$0 = \sum n_k d_k = n_j,$$

as asserted. ■

To illustrate the point that Theorem 10.4 is effectively stronger than Theorem 9.4, we deduce a result which does not follow at once from Theorem 9.4.

Corollary *Let K be homotopy equivalent to a circle. Then $\mathbf{C} - K$ has at most two components.*

Proof. We have $H^1(K) \cong H^1(S^1) \cong \mathbf{Z}$. Then $\tilde{H}_0(\mathbf{C} - K)$ is isomorphic to a subgroup of this, so is either trivial or infinite cyclic. Accordingly, $\mathbf{C} - K$ has one or two path components. ■

FURTHER DEVELOPMENTS

The result mentioned is only the plane case of Alexander's celebrated duality theorem. The general case is that for K a compact subset of $\mathbf{R}^n$, there are duality isomorphisms

$$D_*: H_r(\mathbf{R}^n - K) \to H^{n-r-1}(K)$$

(with slight modification, as above, if $r = 0$ or $n - 1$). Proofs can be found in Alexandroff and Hopf, Lefschetz, Maunder, Pontrjagin, Spanier; these proofs are not related closely to the one we will give.

There is a more general result in which $\mathbf{R}^n$ is replaced by a space locally homeomorphic to $\mathbf{R}^n$ (an n-manifold): this is the Poincaré duality theorem. See the same references for proofs.

Alexandroff, P. and H. Hopf, *Topologie I*, Springer, Berlin, 1935.

Lefschetz, Solomon, *Introduction to Topology*, Princeton University Press, 1949.

Maunder, C. R. F., *Algebraic Topology*, Van Nostrand, Princeton, 1970.

Pontrjagin, L. S., *Foundations of Combinatorial Topology*, (English translation) Graylock, 1952.

Spanier, E. H., *Algebraic Topology*, McGraw-Hill, New York, 1966.

EXERCISES AND PROBLEMS

1. Consider a diagram of six groups and nine homomorphisms, forming a triangular prism as in Lemma 10.2. For which faces A of the prism does the statement, "if all faces but A are commutative, then so is A," hold? If the statement does not hold for A, give an extra hypothesis (e.g. surjectivity or injectivity of some map) under which it does.
2. Let $K \subset L$ be compact subsets of $\mathbf{C}$; D_K and D_L the corresponding duality maps. Show that the diagram

$$\begin{array}{ccc} \tilde{H}_0(\mathbf{C} - L) & \to & \tilde{H}_0(\mathbf{C} - K) \\ \downarrow D_L & & \downarrow D_K \\ H^1(L) & \to & H^1(K) \end{array}$$

(with horizontal maps induced by inclusions) commutes.
3. A *θ-curve* is a space homeomorphic to the union of a circle and one of its diameters. Using Theorem 10.4, show that a θ-curve separates $\mathbf{C}$ into at most three components. Assuming the Jordan curve theorem, show that the number must be exactly 3. Obtain the same results for the disjoint union of two Jordan curves.
4. Let C be a circle, $x \in F(\mathbf{C} - C)$. Show that the degree of $D_3(x)$ on C is the number of points of x (counted with multiplicities) inside C.

5. Give a detailed proof that if V is a connected open subset of $\mathbf{R}^n$, $n \geqslant 2$, and $x \in V$, then $V - \{x\}$ is connected. Where does the argument fail when $n = 1$?
6. What is the kernel of D_2?
7. Adopt the notation of the proof of Theorem 10.4; set $h(z) = (z - a_j)^{-1}$, $h(K) = L$, and U the unbounded component of $\mathbf{C} - L$. Show that there is a continuous map $M : \mathbf{C} \to S^1$ such that

$$M(w) = D_3(x)(a_j + w^{-1}) \qquad \text{for} \quad w \in L \cup U,$$

and hence obtain an alternative proof of Theorem 10.4 by considering the degree of M on circles of large radius with center 0.

11
PROOF OF THE DUALITY THEOREM

INTRODUCTION

We have already mentioned that the duality theorem states that for K a compact subset of $\mathbf{C}$,

$$D: \tilde{H}_0(\mathbf{C} - K) \to H^1(K)$$

is an isomorphism, and have proved D injective. For surjectivity we need a new argument. The most obvious difficulty facing us is the arbitrariness of the set K. Even if K is, for instance, homeomorphic to a circle, it can lie in a very contorted position in the plane. The existence (due to Weierstrass) of a continuous but nowhere differentiable function shows that K need not possess a tangent anywhere. We can illustrate this possibility by an example obtained, like that of Weierstrass, as the limit of a sequence whose nth term is expressible in finite terms. The "snowflake curve" is obtained from an equilateral triangle by repeatedly trisecting all edges and erecting new equilateral triangles whose bases are the middle sections: the desired curve is the outer perimeter of the figure. Figure 11.1 represents the third stage of the construction.

The figure, with the construction lines left in, defines a compact set of considerable complexity. Another, related to the counterexamples of Chapter 3, is the "Hawaiian earring" (Fig. 11.2), defined as the union of a nested sequence of mutually tangent circles, with diameters tending to zero. And finally, Fig. 11.3, the "lakes of Wada," describes a figure formed from a closed disk (the land) and three open subdisks (the lakes) by extending the lakes in turn by canals so that at the nth stage, lake number n (modulo 3) is within a distance $1/n$ of each remaining point of dry land. The limit of the dry land, as $n \to \infty$, gives another compact set with unpleasant properties.

We cannot expect to be able to study in detail the behavior of all such spaces. Thus our proof must effectively replace K by something simpler. This will involve some approximation technique.

In our case, simpler spaces are ready to hand. All we use of $\mathbf{C} - K$ is a finite set of points, defining an element of $H_0(\mathbf{C} - K)$. We cannot enlarge K to all of $\mathbf{C}$ except these points, as this would not be compact. Instead, we

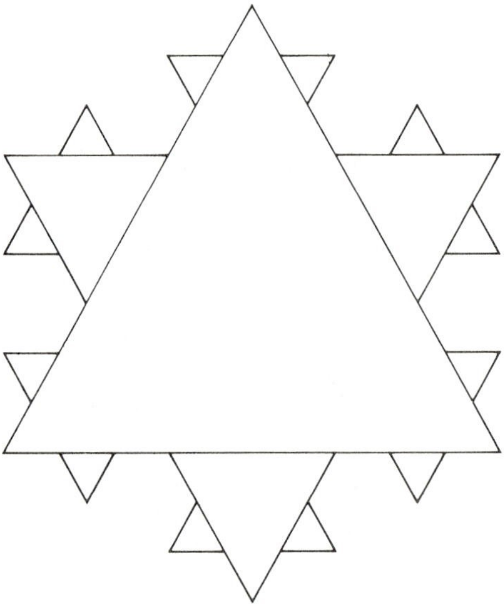

Fig. 11.1

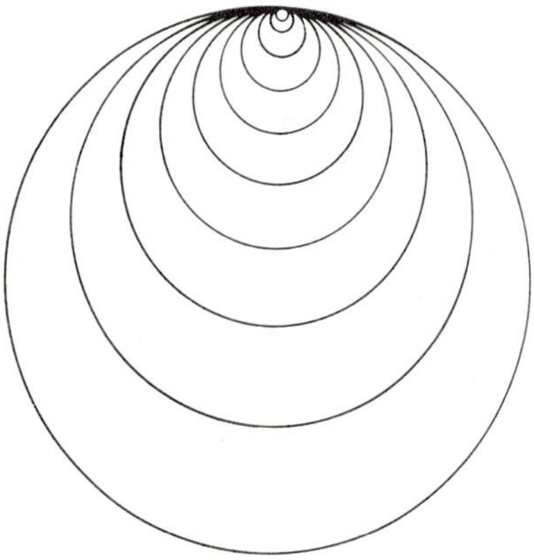

Fig. 11.2

Fig. 11.3

obtain L from a large circular disk, with K and the points in its interior, by deleting small open circular disks centered at the points. L is compact, contains K, and still has the relevant points in its complement. So the idea will be to reduce problems about K to problems about some such L.

AN EXTENSION THEOREM

In trying to prove D surjective, we may suppose given a class ϕ in $H^1(K)$, represented by a map $f: K \to S^1$. The above discussion suggests that we start by extending f to a map $L \to S^1$ for some suitable L. In fact, this extension is the only difficult part of the proof. In the trivial case $\phi = 0$, f is nullhomotopic, so by the corollary to Proposition 7.6, it extends to a continuous map $\mathbf{C} \to S^1$; conversely, since $H^1(\mathbf{C}) = 0$, if such an extension exists, we have $\phi = 0$. In the general case, we have Tietze's extension theorem (7.4). Applying this separately to the real and imaginary parts of f (cf. Exercise 7.2)—the application is justified since the compact set K is closed in $\mathbf{C}$—we have a continuous extension of f to $g: \mathbf{C} \to \mathbf{C}$. Now if 0 is not in the image of g, $N \circ g: \mathbf{C} \to S^1$ extends f, and we are in the trivial case discussed above. In general, we can write $X = g^{-1}(\{0\})$, and

$$N \circ (g|\mathbf{C} - X): \mathbf{C} - X \to S^1$$

then extends f. Thus if X is finite, by restricting $\mathbf{C}$ to a suitably large disk, and removing small disks surrounding the points of X (but disjoint from K), we have an extension of the desired kind. So we next prove

11.1 Theorem *Let $K \subset \mathbf{C}$ be compact, and $f: K \to \mathbf{C} - \{0\}$ continuous. Then there is a continuous extension $g: \mathbf{C} \to \mathbf{C}$ of f, with $g^{-1}(\{0\})$ finite.*

Proof. By Theorem 7.4, we can indeed extend f to a continuous map $g_1: \mathbf{C} \to \mathbf{C}$. We now wish to improve g_1 to achieve the result. The two best established techniques in topology for improving maps are to approximate them by maps which are either differentiable or are linear on each set of some subdivision of the domain. Here we do not care what sort of map g is, nor even whether it approximates g_1; the quickest short cut to the theorem seems to be one using the idea of subdivision and not much more.

First we deal with the noncompactness of $\mathbf{C}$. Let K be contained in $|z| \leqslant r$; write F for the set defined by $|z| \geqslant 2r$. Then the map $f_1: K \cup F \to \mathbf{C}$ defined by $f_1|K = f$, $f_1(F) = \{1\}$ is continuous, and $K \cup F$ is closed in $\mathbf{C}$; by Tietze's theorem, f_1 extends to a continuous map $g_2: \mathbf{C} \to \mathbf{C}$.

Since g_2 is continuous, $X = g_2^{-1}(\{0\})$ is closed; since g_2 extends f_1, X

does not meet F, so X is bounded. Likewise, X does not intersect K. Then by Theorem 1.12, there exists a number $\delta > 0$ with

$$d(z, w) > 2\delta \qquad \text{for all} \quad z \in K, \quad w \in X.$$

Divide the set $|x| \leqslant 2r$, $|y| \leqslant 2r$ by lines parallel to the axes into rectangles with sides of length $\leqslant \delta$. Then no rectangle can intersect both K and X (Fig. 11.4).

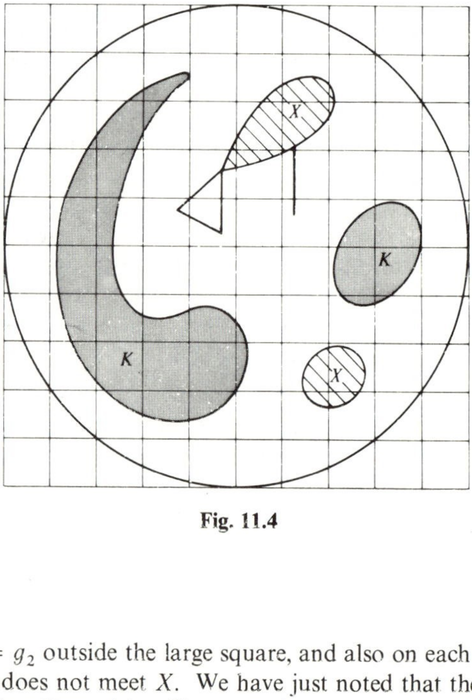

Fig. 11.4

We take $g = g_2$ outside the large square, and also on each rectangle, edge, or corner which does not meet X. We have just noted that the union of these rectangles contains K, so this will ensure that g extends f. For each corner P belonging to X, we define $g(P) = 1$. For each edge PQ which intersects X (perhaps at interior points, or maybe only at an end) we have now already defined $g(P)$ and $g(Q)$ in $\mathbf{C} - \{0\}$. Since $\mathbf{C} - \{0\}$ is path-connected, we can join these values by a path $\gamma: I \to \mathbf{C} - \{0\}$. We now define g on the edge PQ by

$$g((1 - t)P + tQ) = \gamma(t), \qquad 0 \leqslant t \leqslant 1.$$

It remains only to define g on the finite number of rectangles which intersect X. We will do this so that g is continuous on each such rectangle, and

$g^{-1}(\{0\})$ contains only the midpoint of the rectangle. Then $g^{-1}(\{0\})$ will be finite, and Theorem 1.7 shows, by induction, that g will be continuous. Thus the theorem will now follow from the next lemma.

11.2 Lemma *Let R be a rectangle in $\mathbf{C}$; write ∂R for the union of its four sides. Then any continuous map $f: \partial R \to \mathbf{C} - \{0\}$ has a continuous extension $g: R \to \mathbf{C}$ such that $g^{-1}(\{0\})$ consists only of the midpoint of the rectangle.*

Proof. Let z_0 be the midpoint of R (Fig. 11.5). Each point of $R - \{z_0\}$ lies on a unique ray through z_0, which intersects ∂R just once, and so it can be uniquely expressed as

$$z_0 + \lambda(z - z_0), \qquad 0 < \lambda \leqslant 1, \quad z \in \partial R.$$

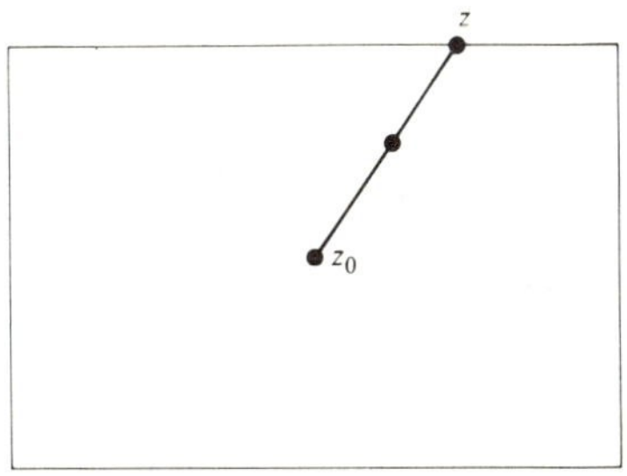

Fig. 11.5

We now define g by

$$g(z_0 + \lambda(z - z_0)) = \lambda f(z), \qquad 0 \leqslant \lambda \leqslant 1, \quad z \in \partial R.$$

At z_0, $\lambda = 0$ and z is not well defined, but the right-hand side vanishes for all $z \in \partial R$. Conversely, since the range of f lies in $\mathbf{C} - \{0\}$, the right-hand side vanishes only for $\lambda = 0$, that is, at z_0. Evidently, g extends f (take $\lambda = 1$).

It remains to show that g is continuous. At a point P of R other than z_0, λ and z depend continuously on P, hence so does $g(P) = \lambda f(z)$. Now since ∂R is compact, $f(\partial R)$ is bounded, say by K. Then if $2a$ is the length of the shorter side of R, and $|P - z_0| < a\varepsilon/K$, then the parameter λ (for P) is less than ε/K, and so $|g(P)| < \varepsilon$, proving continuity of g at z_0. ■

NATURALITY

The following simple result relates duality maps for different subsets of $\mathbf{C}$.

11.3 Lemma *Let K, L be (compact) subsets of $\mathbf{C}$ with $K \subset L$; denote the inclusion maps by $i_1 : K \subset L$ and $i_2 : \mathbf{C} - L \subset \mathbf{C} - K$. Then the following diagram is commutative:*

$$\begin{array}{ccc} \tilde{H}_0(\mathbf{C} - L) & \xrightarrow{i_{2*}} & \tilde{H}_0(\mathbf{C} - K) \\ \downarrow D^L & & \downarrow D^K \\ H^1(L) & \xrightarrow{i_1^*} & H^1(K). \end{array}$$

Proof. First observe that

$$\begin{array}{ccc} \mathbf{C} - L & \xrightarrow{i_2} & \mathbf{C} - K \\ \downarrow D_0^L & & \downarrow D_0^K \\ \text{Map}\,(L, S^1) & \xrightarrow{i_1^*} & \text{Map}\,(K, S^1) \end{array}$$

commutes; indeed for $a \in \mathbf{C} - L$ and $z \in K$, we have

$$D_0^K(i_2(a))(z) = D_0^K(a)(z) = N(z - a) = D_0^L(a)(z).$$

The corresponding diagram with $F(\mathbf{C} - L)$ and D_3 now commutes, as we see by using the universal property. The result follows on passing to homotopy classes. ■

PROOF IN SOME SPECIAL CASES

First suppose K is a circle $|z - z_0| = r_0$. Then $\mathbf{C} - K$ has just two components:

$$\begin{array}{ll} B \text{ (bounded)}, & |z - z_0| < r_0, \\ U \text{ (unbounded)}, & |z - z_0| > r_0 \end{array}$$

(these are both open and path-connected, as we have seen earlier); this computes $\tilde{H}_0(\mathbf{C} - K)$. We compute $H^1(K)$ by noting that there is a homeomorphism $h: S^1 \to K$ defined by

$$h(z) = z_0 + r_0 z.$$

Since h is a homeomorphism, $h^*: H^1(K) \to H^1(S^1)$ is an isomorphism; since $\deg: H^1(S^1) \to \mathbf{Z}$ is an isomorphism, so is $d_0 = \deg \circ h^*$. We will also write $d_0(f)$ for $f: K \to S^1$.

We now compute D_1. By the above, it is more explicit to compute $d_0 \circ D_1$. By Lemma 11.3, $D_1(U) = 0$. But $B = p(z_0)$, so

$$d_0(D_1(B)) = d_0(D_1(p(z_0))) = d_0(D_0(z_0)) = \deg\left(D_0(z_0) \circ h\right).$$

But

$$\begin{aligned} D_0(z_0)(h(z)) &= D_0(z_0)(z_0 + r_0 z) \\ &= N(r_0 z) \\ &= z, \end{aligned}$$

so the map is the identity map and the degree is 1. The elements of $\tilde{H}_0(\mathbf{C} - K)$ are the $mi'(B) + ni'(U)$ with $m + n = 0$, that is, $n = -m$; and by the above, $d_0 \circ D$ maps this to $m \in \mathbf{Z}$. Thus we have a bijection, hence an isomorphism.

We use this in the next case, which is the one we will apply.

Let Δ_0 be the disk $|z - z_0| \leqslant r_0$ in $\mathbf{C}$, and suppose that the disks Δ_i given by $|z - z_i| \leqslant r_i$ $(1 \leqslant i \leqslant k)$ are disjoint, and lie in the interior of Δ_0. Analytically this means

$$\begin{aligned} |z_i - z_0| &< r_0 - r_i, \qquad 1 \leqslant i \leqslant k, \\ |z_i - z_j| &> r_i + r_j, \qquad 1 \leqslant i < j \leqslant k. \end{aligned}$$

Let X be the space obtained from Δ_0 by deleting the interiors of the disks Δ_i; thus X is defined by

$$|z - z_0| \leqslant r_0, \qquad |z - z_i| \geqslant r_i, \qquad 1 \leqslant i \leqslant k$$

(Fig. 11.6). Write Γ_i for the circle $|z - z_i| = r_i$ $(0 \leqslant i \leqslant k)$, and Y for the union of the Γ_i with $1 \leqslant i \leqslant k$; let $j\colon Y \subset X$.

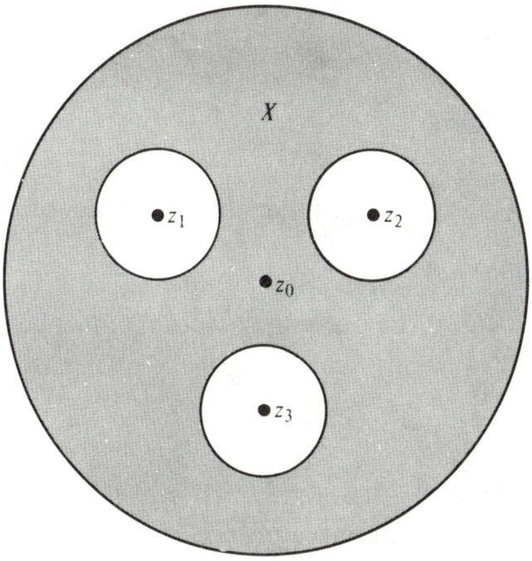

Fig. 11.6

11.4 Lemma *$j^*: H^1(X) \to H^1(Y)$ is injective.*

Proof. Let $f: X \to S^1$ be such that $f|Y$ is nullhomotopic. Then for each i $(1 \leqslant i \leqslant k)$, $f|\Gamma_i$ is nullhomotopic. By the corollary to Proposition 7.6, it extends to a continuous map $g_i: \Delta_i \to S^1$. If we now define $F: \Delta_0 \to S^1$ by $F|X = f$, $F|\Delta_i = g_i$ $(1 \leqslant i \leqslant k)$, it follows from Theorem 1.7 that F is continuous.

But Δ_0 is convex and so contractible. Thus F is nullhomotopic, and so by Lemma 5.2 is $f = F|X$. Thus j^* has zero kernel, so by Lemma 2.1 it is injective. ■

11.5 Proposition *$j^* \circ D: \tilde{H}_0(\mathbf{C} - X) \to H^1(Y)$ is an isomorphism.*

Note that it follows from this that D is an isomorphism, i.e. that the duality theorem holds for X. For if $D(x) = 0$, then $j^*(D(x)) = 0$; since $j^* \circ D$ is injective, $x = 0$. Thus D is injective. On the other hand, if $y \in H^1(X)$, then since $j^* \circ D$ is surjective, there exists $z \in \tilde{H}_0(\mathbf{C} - X)$ with $j^*(D(z)) = j^*(y)$. Since j^* is injective, we obtain $y = D(z)$. Thus D is surjective.

Proof. As in the case above (K a circle), it is clear that $\mathbf{C} - X$ has components $E = \mathbf{C} - \Delta_0$ and $\operatorname{Int} \Delta_i$ $(1 \leqslant i \leqslant k)$. Then any element of $H_0(\mathbf{C} - X) = F(\pi_0(\mathbf{C} - X))$ can be uniquely written as

$$n_0 i'(\mathbf{C} - \Delta_0) + \sum_{j=1}^{k} n_j i'(\operatorname{Int} \Delta_j).$$

This belongs to $\tilde{H}_0(\mathbf{C} - X)$ if and only if

$$n_0 + \sum_{j=1}^{k} n_j = 0,$$

that is,

$$n_0 = -\sum_{j=1}^{k} n_j;$$

so we can regard the n_j $(1 \leqslant j \leqslant k)$ as independent integers and n_0 as determined by them.

Now by Lemma 8.2, $H^1(Y)$ is the direct sum of the $H^1(\Gamma_j)$ $(1 \leqslant j \leqslant k)$, and as in the preceding example by using the homeomorphism $h_j: S^1 \to \Gamma_j$ given by

$$h_j(z) = z_j + r_j z,$$

we define an isomorphism $d_j = \deg \circ h_j^*: H^1(\Gamma_j) \to \mathbf{Z}$. Thus elements of $H^1(Y)$ correspond bijectively to sequences of integers $(d_1, \ldots, d_k)$.

It remains to compute the map D. It is enough to determine the numbers

$$d_j\left(D_2\left(n_0 i'(\mathbf{C} - \Delta_0) + \sum_{j=1}^{k} n_j i'(\operatorname{Int} \Delta_j)\right)\right).$$

Consider the diagram

$$\begin{array}{ccccc} H_0(\mathbf{C} - X) & \rightarrow & H_0(\mathbf{C} - \Gamma_j) & & \\ \downarrow{\scriptstyle D_2(K)} & & \downarrow{\scriptstyle D_2(\Gamma_j)} & & \\ H^1(X) \xrightarrow{j^*} H^1(Y) & \rightarrow & H^1(\Gamma_j) & \xrightarrow{d_j} & \mathbf{Z} \end{array}$$

by Lemma 11.3, this is commutative. But we computed $D_2(\Gamma_j)$ above, and Int Δ_j (corresponding to B above) is the bounded component of $\mathbf{C} - \Gamma_j$ $(1 \leqslant j \leqslant k)$. Thus d_j is the coefficient of Int Δ_j, and so $d_j = n_j$. It now follows from the above descriptions of $\tilde{H}_0(\mathbf{C} - X)$ and $H^1(Y)$ that $j^* \circ D$ is bijective. ■

END OF THE PROOF

Recall the statement of the result.

11.6 Theorem *If K is a compact subset of $\mathbf{C}$, the map*

$$D\colon \tilde{H}_0(\mathbf{C} - K) \rightarrow H^1(K)$$

is an isomorphism.

Proof. We saw in Theorem 10.4 that D is injective: now we must prove it surjective. Let $\phi \in H^1(K)$; represent ϕ by a map $f\colon K \rightarrow S^1$. By Theorem 11.1, we can extend f to a map $g\colon \mathbf{C} \rightarrow \mathbf{C}$ with $g^{-1}(0)$ finite. Set $z_0 = 0$ and choose r_0 so large that $K \subset U(0, r_0)$; let $z_1, \ldots, z_k$ be the points of $g^{-1}(0)$ in $U(0, r_0)$ and choose numbers r_i so small that $U(z_i, r_i) \subset \mathbf{C} - K$; thus if X is defined as above, $K \subset X$. Now $g|X$ has image in $\mathbf{C} - \{0\}$, so $N \circ (g|X)\colon X \rightarrow S^1$ extends f. Thus in the commutative diagram of Lemma 11.3

$$\begin{array}{ccc} \tilde{H}_0(\mathbf{C} - X) & \xrightarrow{i_{2*}} & \tilde{H}_0(\mathbf{C} - K) \\ \downarrow{\scriptstyle D^X} & & \downarrow{\scriptstyle D^K} \\ H^1(X) & \xrightarrow{i_1^*} & H^1(K) \end{array}$$

ϕ is in the image of i_1^*. But since according to Proposition 11.5 the theorem holds for X, D^X is surjective. So ϕ is in the image of

$$i_1^* \circ D^X = D^K \circ i_{2*},$$

and hence in that of D^K. But ϕ was arbitrary, so D^K is indeed surjective. This completes the proof. ■

The Jordan curve theorem is an immediate consequence.

11.7 Theorem *Let $K \subset \mathbf{C}$ be homeomorphic to S^1. Then $\mathbf{C} - K$ has exactly two path components.*

Since K is homeomorphic to S^1, it is compact, so Theorem 11.6 is applicable. Hence

$$\tilde{H}_0(\mathbf{C} - K) \cong H^1(K) \cong H^1(S^1) \cong \mathbf{Z},$$

and

$$H_0(\mathbf{C} - K) \cong \tilde{H}_0(\mathbf{C} - K) \oplus \mathbf{Z} \cong \mathbf{Z} \oplus \mathbf{Z}$$

has rank 2. So $\mathbf{C} - K$ has two path components, as asserted. ■

FURTHER DEVELOPMENTS

We have already mentioned extensions to higher dimensions. As described above, the key idea of the proof is to replace an arbitrary compact set K by something "close enough" and "reasonably nice." There are two standard interpretations of the second requirement: we may seek a subset which is either polyhedral or differentiable (we have used the second, but the choice is purely a matter of taste). It can be shown that if U is any open set containing K, then there exists a compact set L, $K \subset L \subset U$ with L polyhedral, and a similar L which is differentiable. For the proofs of these assertions, see Stallings or Hudson, and Conner and Floyd or Stong.

Conner, P. E. and E. E. Floyd, *Differentiable Periodic Maps*, Springer, Berlin, 1964.

Hudson, J. F. P., *Piecewise Linear Topology*, Benjamin, New York, 1969.

Stallings, J. R., *Lectures on Polyhedral Topology*, Tata Institute, 1968.

Stong, R. E., *Notes on Cobordism Theory*, Princeton University Press, 1968.

EXERCISES AND PROBLEMS

1. Construct an extension $g: D^2 \to \mathbf{C}$ of the identity map of S^1 such that $g^{-1}(\{0\})$ is not finite.
2. Let $g: \mathbf{C} \to \mathbf{C}$ extend $f: K \to S^1$ ($K \subset \mathbf{C}$ compact), and suppose f homotopic to f'. Show that there is an extension $g': \mathbf{C} \to \mathbf{C}$ of f' with $g'^{-1}(\{0\}) = g^{-1}(\{0\})$. [*Hint:* One can choose g' equal to g except near K.]
3. Let $L \subset \mathbf{C}$ be a connected graph with α_0 vertices and α_1 edges; let $\mathbf{C} - L$ have α_2 components. Prove Euler's formula $\alpha_0 - \alpha_1 + \alpha_2 = 2$.
4. Show that if K_1 and K_2 are compact subsets of $\mathbf{C}$ such that K_1 separates $\mathbf{C}$ and K_2 dominates K_1, then K_2 separates $\mathbf{C}$.

6. Show that if X is as in Lemma 11.4 and $f: X \times S^1$ has degree d_i on Γ_i $(0 \leqslant i \leqslant k)$, then

$$d_0 = \sum_{i=1}^{k} d_i.$$

7. Show that if $K \subset D^2$ is compact, and $g: D^2 \to \mathbf{C}$ is a map with $g(K) \subset \mathbf{C} - \{0\}$, we can find a map $g': D^2 \to \mathbf{C}$, agreeing with g on K, having $g'^{-1}(\{0\})$ finite, and satisfying $|g'(z) - g(z)| < \varepsilon$ for all z, where $\varepsilon > 0$ is any given number. [*Hint:* First choose

$$\eta < \min\left(\frac{\varepsilon}{3}, d(g(K), 0)\right).$$

Then chop D^2 up by rectangles so small that none meets K and $g^{-1}(U_\eta(0))$, and that

$$|g(z_1) - g(z_2)| < \eta$$

for z_1 and z_2 in the same rectangle. Then argue as for Theorem 11.1.]

8. Show that Exercise 10.4 is true if C is any Jordan curve; in particular, that if C separates P from Q, then

$$D_3(i(P) - i(Q)) \circ h$$

has degree ± 1 for any homeomorphism h of S^1 on C.

9. Let

$$A_i = (i, 0, 0), \qquad B_i = (0, i, 1), \qquad i = 1, 2, 3$$

in $\mathbf{R}^3$. Let G be the union of the nine segments $A_i B_j$ $(i, j = 1, 2, 3)$. List all the Jordan curves contained in G. List all elements of $H^1(G)$ whose restriction to each of these Jordan curves has degree $+1$, 0, or -1.

10. Using the preceding two exercises, show that there is no subspace of $\mathbf{C}$ homeomorphic to G.

11. Discuss the groups $H_0(\mathbf{C} - K)$ and $H^1(K)$ for the compact spaces K described at the beginning of the chapter.

12
REMARKS ON THE PROOF

INTRODUCTION

I have tried to describe the proof of the duality theorem in the preceding three chapters in the simplest possible terms, since an understanding of basic results in topology does not really depend on much other mathematics. However, if we look more closely at the proof, we find rich associations with other branches of mathematics, which make our constructions and arguments seem more natural, and give some insight into why the proof works.

THE EXTENDED PLANE

We first observe that it is more aesthetically satisfactory to close up, or more accurately, compactify the plane **C** by adding an extra point. This can be done as follows. Take the unit sphere S^2 in $\mathbf{R}^3$ and project stereographically from the north pole (0, 0, 1) onto the equatorial plane $x_3 = 0$, which we can identify (by $z = x_1 + ix_2$) with **C** (Fig. 12.1). The map, which is well defined everywhere except at (0, 0, 1) itself, is expressed in terms of coordinates by

$$z = \frac{x_1 + ix_2}{1 - x_3};$$

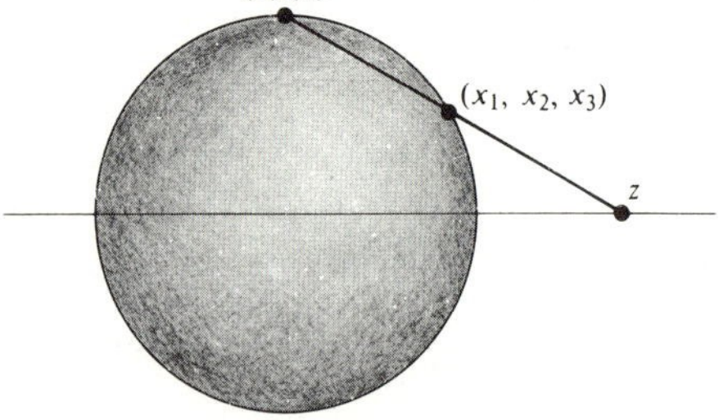

Fig. 12.1

except at $(0, 0, -1)$ we have also the alternative expression $(1 + x_3)/(x_1 - ix_2)$. There is a two-sided inverse map, given by

$$(x_1, x_2, x_3) = \left(\frac{2 \operatorname{Re} z}{|z|^2 + 1}, \frac{2 \operatorname{Im} z}{|z|^2 + 1}, \frac{|z|^2 - 1}{|z|^2 + 1}\right) = e(z), \quad \text{say.}$$

Since each of these maps is evidently continuous, we have constructed a homeomorphism e between $\mathbf{C}$ and $S^2 - \{(0, 0, 1)\}$.

It is convenient to regard S^2 as obtained from $\mathbf{C}$ by adjoining this point; accordingly, we regard z as a coordinate on S^2, and assign to $(0, 0, 1)$ the symbolic coordinate $z = \infty$. Readers familiar with projective geometry will recognize the process of adjoining a point at infinity to $\mathbf{C}$ (the "complex affine line") to obtain a one-dimensional complex projective space; here, however, topological rather than projective properties are in question.

In order to study continuity on S^2 with respect to these complex coordinates, note that (by the above) z is a good coordinate except at $(0, 0, 1)$. Now the reflection of S^2 in the x_1-axis is a homeomorphism expressed in complex coordinates by $z \to z^{-1}$ (with the familiar conventions $\infty = 0^{-1}$, $0 = \infty^{-1}$). Thus z^{-1} is an appropriate coordinate for investigating the topology of the extended plane at ∞.

We now see (using, for example, Lemma 3.5(iii)) that S^2 is l.p.c., and so Lemmas 9.1 and 9.2 apply with $\mathbf{C}$ replaced by S^2. (However, Lemma 9.3 does not, for S^2 is compact, and the notion of boundedness does not arise.) Moreover, the separation properties of $\mathbf{C}$ and S^2 are essentially the same. For if K is a subset of S^2, then either $K = S^2$, in which case nothing is left to study, or there is a point $P \in S^2 - K$, and we can then rotate S^2 to make $P \mapsto \infty$, and the image of K is then contained in that of $\mathbf{C}$. For this case we have

12.1 Lemma *Let K be a compact subset of* $\mathbf{C}$; *denote by B_α the bounded components of $\mathbf{C} - K$ and by U the unbounded. Then the components of $S^2 - e(K)$ are the $e(B_\alpha)$ and $e(U) \cup \{\infty\}$.*

Proof. As in Lemma 9.3, let K be contained in $|z| \leqslant r$, and define E by $|z| > r$. Then $e(E) \cup \{\infty\}$ is a disk in S^2, and so is connected; it also contains ∞ in its interior. Hence $e(U) \cup \{\infty\}$, the union of the intersecting connected sets $e(U)$ and $e(E) \cup \{\infty\}$, is connected.

We will now show that each of the sets $e(B_\alpha)$, $e(U) \cup \{\infty\}$ is open in S^2 [and hence in $S^2 - e(K)$]. It then follows that they are the path components (= components) of $S^2 - e(K)$, since any two of them are separated by some splitting, so no path in $S^2 - e(K)$ can join two of them.

Since B_α, U are open in $\mathbf{C}$, and $e(\mathbf{C})$ is open in S^2, $e(B_\alpha)$ and $e(U)$ are open in S^2. But we have already seen that ∞ is an interior point of $e(U) \cup \{\infty\}$. Hence each of $e(B_\alpha)$, $e(U) \cup \{\infty\}$ is open in S^2. ■

REFORMULATION OF PRECEDING CHAPTERS

In projective geometry it is usual to use homogeneous coordinates (z_0, z_1), where $z = z_0/z_1$ was the original coordinate on the line, and $z = \infty$ corresponds to $z_0 = 0$. We must agree that z_0 and z_1 may not vanish simultaneously, and that for any $\lambda \in \mathbf{C} - \{0\}$, $(\lambda z_0, \lambda z_1)$ denotes the same point as (z_0, z_1).

One now considers the "bilinear" transformations

$$\left.\begin{aligned} z_0' &= az_0 + bz_1 \\ z_1' &= cz_0 + dz_1 \end{aligned}\right\} \qquad \text{subject to} \qquad 0 \neq \begin{vmatrix} a & b \\ c & d \end{vmatrix} = ad - bc.$$

This is well defined and invertible, and the induced map is called a projective map. In the old coordinates, it was given by

$$z \mapsto z' = \frac{az + b}{cz + d}.$$

It is essential for us to observe that this defines a continuous map $S^2 \to S^2$. Continuity is clear except, perhaps, at $z = \infty$ and $z = -d/c$ (which coincide if $c = 0$) but follows there on making the substitution $w = z^{-1}$. The projective maps form a group, the projective group $PGL_2(\mathbf{C})$, acting on S^2 by homeomorphisms.

12.2 Lemma (*Fundamental Theorem of Projective Geometry*). *The operation of this group is sharply triply transitive.*

In other words, given two sets each of three distinct points (P_1, P_2, P_3) and (Q_1, Q_2, Q_3), there is a unique projective map f with $f(P_i) = Q_i$ for $i = 1, 2, 3$.

We will not use this result, so do not prove it in detail. In the case $(Q_1, Q_2, Q_3) = (0, 1, \infty)$, the desired map is given (in inhomogeneous coordinates) by

$$z \mapsto (z, P_2; P_1, P_3) = \frac{(z - P_1)(P_2 - P_3)}{(z - P_3)(P_2 - P_1)}. \quad \blacksquare$$

The symbol $(z, P_2; P_1, P_3)$ just defined is called the *cross ratio* of the four points. The notion of cross ratio plays the role in projective geometry that distance does for euclidean geometry. The lemma shows, in particular, that we can require two distinct points a, b to be mapped to 0 and ∞; if a, b are finite, this can of course be achieved by

$$z \to \frac{z - a}{z - b}.$$

Thus, using Lemma 12.1, we can restate Eilenberg's criterion as:

12.3 Lemma *Let K be a compact subset of S^2 not containing 0 or ∞. Then K separates 0 from ∞ if, and only if, $N \circ (e^{-1}|K): K \to S^1$ is not nullhomotopic.* ∎

Next, we consider the construction of the duality map. The basic definition in Chapter 10 was

$$D_0(a)(z) = N(z - a), \qquad a \in \mathbf{C} - K, \quad z \in K.$$

In homogeneous coordinates, this becomes $N(z_0 a_1 - z_1 a_0)$, which is not well defined. The expression

$$N\left(\frac{z - a}{z - b}\right)$$

occurring in Theorem 9.4 seems more significant; here we must rewrite it as

$$N\left(\frac{z_0 a_1 - z_1 a_0}{z_0 b_1 - z_1 b_0}\right).$$

Even this depends on the normalization of the homogeneous coordinates of a and b, so we are led to consider the cross ratio $(z, w: a, b)$. As we do not wish this to vanish for $z \in K$, $a \in S^2 - K$, it is natural to take $w \in K$ and $b \in S^2 - K$, and to hold these fixed: we call them base points. Now ∞ has no special role, and we define $D'_0: S^2 - K \to \text{Map}(K, S^1)$ by

$$D'_0(a)(z) = N(z, w; a, b).$$

The definition of further maps D'_1, D'_2, D'_3 and the construction of a diagram analogous to Lemma 10.2 now proceeds exactly as in Chapter 10. In particular,

$$\begin{aligned} D'_3\left(\sum n_j i(a_j)\right)(z) &= N\left(\prod_j \left\{\left(\frac{z_0 a_{j1} - z_1 a_{j0}}{z_0 b_1 - z_1 b_0}\right)\left(\frac{w_0 b_1 - w_1 b_0}{w_0 a_{j1} - w_1 a_{j0}}\right)\right\}^{n_j}\right) \\ &= N\left(\prod_j \left(\frac{z_0 a_{j1} - z_1 a_{j0}}{w_0 a_{j1} - w_1 a_{j0}}\right)^{n_j} \cdot \left(\frac{w_0 b_1 - w_1 b_0}{z_0 b_1 - z_1 b_0}\right)^{\Sigma n_j}\right). \end{aligned}$$

The case $\sum n_j = 0$ corresponding to $\tilde{H}_0(S^2 - K)$ now gives rise to the simplification that the second term disappears, so the expression is independent of b.

Though it still depends on w, this only gives a constant factor, which does not affect homotopy class, so that

$$D': \tilde{H}_0(S^2 - K) \to H^1(K)$$

does not depend on either choice of base point; also on $\mathbf{C} - K$ it is given by essentially the same formula as D.

With these definitions, Lemma 10.3 is replaced by the observation that D_2 annihilates the component of $S^2 - K$ containing the base point b, and that if b is taken as ∞, with homogeneous coordinate (0, 1), we recover the formulas of Chapter 10 up to a constant factor depending on w.

The proof of Theorem 10.4 is also simplified, for ∞ now plays no special role. If $D_3'(x)$ is nullhomotopic on K, we can extend it to all components A_j of $S^2 - K$ except one by the corollary to Proposition 7.6, and to all points in A_j except a_j as $D_3'(x)$. If now a small disk D_j round a_j, with boundary C_j is deleted, the complement is a disk, hence contractible, so $D_3'(x)|C_j$ is nullhomotopic, and hence of degree zero. Arguing on the other hand as before, we find the degree to be $\pm n_j$ (the sign depending on the chosen homeomorphism of S^1 on C_j). This argument depends on the absence of a further exceptional point b, and hence on the fact that we are dealing directly with $\tilde{H}_0(S^2 - K)$.

The replacement of $\mathbf{C}$ by S^2 affects Chapter 11 less than Chapter 10; indeed, the given proof of Theorem 11.1 shows that for K a compact subset of $e(\mathbf{C})$, any continuous $f: K \to S^1$ extends to a continuous $g: S^2 \to \mathbf{C}$ with $g^{-1}(\{0\})$ a finite set $z_1, \ldots, z_r$. There is now no need to mention D_0; simply remove from S^2 the interiors of disks D_j around z_j, disjoint from K and from each other. Call the result Y_r. Then f is extended by $h = N \circ (g|Y_r): Y_r \to S^1$, with degree n_j (say) on the boundary circle C_j of D_j. Necessarily (see below), $\sum n_j = 0$, and now $h \simeq D_3'(\sum n_j i(z_j))$, essentially by Lemma 11.5.

Note that $D_3'(\sum n_j i(z_j)) = N(F)$, where F is a rational function on S^2, having zeros at the a_j with multiplicities n_j (these are really poles if $n_j < 0$) and (if $\sum n_j = 0$, so there is no base point b to consider) no more. This property determines F up to a multiplicative constant. For rational functions on S^2 (as opposed to $\mathbf{C}$), the sum of the multiplicities of all zeros (including poles) vanishes, which justifies the assertion $\sum n_j = 0$ above. This is yet another reason why we restricted the duality map from H_0 to $\tilde{H}_0$.

THE HOPF MAP

We conclude this chapter by reexamining in detail the relation between our complex homogeneous coordinates on the sphere S^2, and ordinary three-dimensional euclidean coordinates. Homogeneous coordinates arise as follows. A projective line is defined as the set of lines through the origin in a two-dimensional vector space. Here, the vector space is taken to be $\mathbf{C}^2$: each point (z_0, z_1) other than (0, 0) lies on a unique line through the origin; the other nonzero points on the line are the $(\lambda z_0, \lambda z_1)$ with $\lambda \in \mathbf{C} - \{0\}$. Thus a point of the projective space is a line in $\mathbf{C}^2$, and is represented by the coordinates (z_0, z_1) of any point on it.

The topologist (who prefers compact sets to general ones) notices that $\mathbf{C}^2$ can be identified with $\mathbf{R}^4$, and thus contains a unit 3-sphere S^3, given by

$$|z_0|^2 + |z_1|^2 = 1.$$

Further, every line through the origin [represented by (z_0, z_1)] intersects this sphere; for example, in $(\lambda z_0, \lambda z_1)$ with $\lambda = (|z_0|^2 + |z_1^2|)^{-1/2}$. Thus the homogeneous coordinates can be restricted to satisfy $|z_0|^2 + |z_1|^2 = 1$. They then determine a point of S^3. Now these are not coordinates in the usual sense: each value of them gives a unique point of S^2, but the coordinates of the point are not unique. In other words, we have defined a function $H: S^3 \to S^2$. We next express this function in our coordinates. The point

$$(z_0, z_1) \in S^3 \subset \mathbf{C}^2$$

with $|z_0|^2 + |z_1|^2 = 1$ becomes the point of S^2 with these homogeneous coordinates, so with the single complex coordinate, $z = z_0/z_1$. We have seen that in terms of euclidean coordinates we have

$$\frac{z_0}{z_1} = z = \frac{x_1 + ix_2}{1 - x_3} = \frac{1 + x_3}{x_1 - ix_2},$$

and that solving this (using $|z_0|^2 + |z_1|^2 = 1$) gives

$$x_1 + ix_2 = \frac{2z}{|z|^2 + 1} = 2z_0\bar{z}_1, \qquad x_3 = \frac{|z|^2 - 1}{|z|^2 + 1} = |z_0|^2 - |z_1|^2.$$

It follows that H is a continuous map. It was first constructed by H. Hopf, and is known as the Hopf map.

For each point of S^2, the homogeneous coordinate (z_0, z_1) is nonunique in that it can be replaced by $(\lambda z_0, \lambda z_1)$ for any λ such that

$$1 = |\lambda z_0|^2 + |\lambda z_1|^2 = |\lambda|^2 |z_0|^2 + |\lambda|^2 |z_1|^2 = |\lambda|^2,$$

that is, such that $\lambda \in S^1$. In fact, this nonuniqueness can be uniformly described. Write $\mathbf{C}_1$ for $S^2 - \{\infty\}$, $\mathbf{C}_2$ for $S^2 - \{0\}$ admitting coordinates z and $w = z^{-1}$ respectively.

12.4 Theorem *For each $i = 1, 2$ there is a homeomorphism*

$$h_i: \mathbf{C}_i \times S^1 \to H^{-1}(\mathbf{C}_i)$$

such that $H \circ h_i$ is the projection on the first factor.

Proof. We simply give formulas defining the h_i, and leave the reader to verify the properties asserted:

$$h_1(z, e^{i\theta}) = \left(\frac{ze^{i\theta}}{\sqrt{1 + |z|^2}}, \frac{e^{i\theta}}{\sqrt{1 + |z|^2}}\right),$$

and, writing $w = z^{-1}$, we have

$$h_2(w, e^{i\theta}) = \left(\frac{e^{i\theta}}{\sqrt{|w|^2 + 1}}, \frac{we^{i\theta}}{\sqrt{|w|^2 + 1}}\right). \blacksquare$$

Note that h_1 and h_2 do not agree on the intersection of their domains: the second formula must be multiplied by $N(z)$ to give the first. Thus we have not given a homeomorphism $S^2 \times S^1 \to S^3$. In fact, no such homeomorphism exists, for

$$H^1(S^2 \times S^1) \cong \mathbf{Z}, \qquad H^1(S^3) = 0.$$

A map like H which is "locally" the projection map of a product is called the projection map of a *fibre bundle*, and local product maps like h_1, h_2 are called *charts* of it.

The images of h_1 and h_2 omit respectively the circles $z_2 = 0$, $z_1 = 0$ in S^3. If S^3 minus a point is represented stereographically on $\mathbf{R}^3$, it is seen that these two circles are simply linked together. The same holds for the preimages by H^{-1} of any two distinct points of S^2; this property is important in the deeper investigation of the topology of the map.

FURTHER DEVELOPMENTS

For more about fibre bundles and the Hopf map, see Steenrod's classic book; or, for example, Husemoller.

Husemoller, Dale, *Fibre Bundles*, McGraw-Hill, New York, 1966.

Steenrod, N. E., *Topology of Fibre Bundles*, Princeton University Press, 1951.

EXERCISES AND PROBLEMS

1. Let a, b, c, d be complex numbers, not all 0. Write out a proof that the map $S^2 \to S^2$ defined by

$$z \mapsto \frac{az + b}{cz + d}$$

is continuous.

2. Write out a full proof of Lemma 12.2, using the hint given in the text.
3. Let K be a compact subset of S^2; a, b distinct points not in K; $h: S^2 \to S^2$ a projective map with $a, b \to 0, \infty$. Show that the homotopy class of $N \circ h: K \to S^1$ depends only on a and b, *not* on the choice of h.

4. Show that each rational function $f(z)$ of z defines a continuous map $\rho: S^2 \to S^2$, and that the number (deg ρ) of solutions of $\rho(x) = a$ (counted with appropriate multiplicities) is independent of a. Show that the number of solutions of $\rho(x) = x$ (properly counted) is $1 + \deg \rho$.

*5. Prove a result analogous to Theorem 12.4 for the map $S^{2n+1} \to P_n(\mathbf{C})$ [where $P_n(\mathbf{C})$ is n-dimensional complex projective space] defined using homogeneous coordinates.

*6. Show that the corresponding map for real projective space, $S^n \to P_n(\mathbf{R})$, is a covering map in the sense of Exercise 6.9. Deduce that there are just two homotopy classes of maps $S^1 \to P_2(\mathbf{R})$.

7. Show that the map $f: S^2 \to \mathbf{R}^6$ defined by

$$f(x, y, z) = (x^2, y^2, z^2, \sqrt{2}yz, \sqrt{2}zx, \sqrt{2}xy)$$

is 2 to 1, and deduce that f defines an embedding of $P_2(\mathbf{R})$. Show also that the image lies in a 4-dimensional sphere.

*8. Consider S^3 as the group of unit quaternions. Each unit quaternion q defines a rotation of the 3-dimensional space of quaternions of trace 0 by $x \mapsto x^q = q^{-1}xq$. Show that this induces a homeomorphism of $P_3(\mathbf{R})$ on SO_3.

9. Construct an embedding of $P_2(\mathbf{C})$ in a euclidean space by the method of Exercise 7, but using such products as $\bar{x}y$.

PART 3
FURTHER RESULTS IN THE TOPOLOGY OF PLANE SETS

13
THE JORDAN CURVE THEOREM

INTRODUCTION

A subset of the plane homeomorphic to S^1 is called a Jordan curve. The Jordan curve theorem, first stated by Camille Jordan, was in fact first correctly proved by Veblen in 1905. We have already shown in Lemma 11.4 that the result follows from the general duality theorem. In this chapter we give two other proofs, the first based on Eilenberg's criterion and the second quite different. We then go on to deduce some classical theorems of plane topology.

THETA CURVES

The idea of the first proof is to break the Jordan curve into two pieces, and then to consider the effect of varying one of the pieces. Ultimately the curve is replaced by a circle or semicircle, for which the result is clear. The first lemma shows how this affects degrees.

Let Θ be the union of S^1 with the interval $[-1, 1]$ on the real axis. Then Θ contains three Jordan curves: S^1 itself, an upper one U, and a lower one L (Fig. 13.1). Homeomorphisms $e'_1 : S^1 \to U$ and $e'_2 : S^1 \to L$ are given by

$$e'_1(x+iy) = \begin{cases} x+iy, & x^2+y^2=1, \quad y \geqslant 0, \\ x, & x^2+y^2=1, \quad y \leqslant 0; \end{cases}$$

$$e'_2(x+iy) = \begin{cases} x, & x^2+y^2=1, \quad y \geqslant 0, \\ x+iy, & x^2+y^2=1, \quad y \leqslant 0. \end{cases}$$

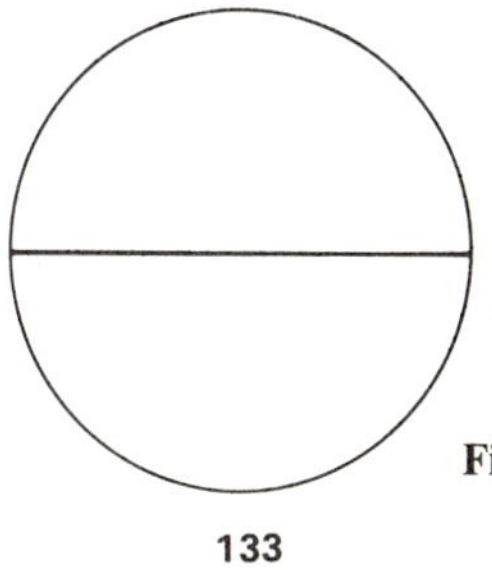

Fig. 13.1

Write $e_0: S^1 \to \Theta$ for the inclusion, and $e_1, e_2: S^1 \to \Theta$ for the maps defined by e'_1, e'_2.

13.1 Lemma *For any* $f: \Theta \to S^1$,

$$\deg(f \circ e_0) = \deg(f \circ e_1) + \deg(f \circ e_2).$$

Proof. Let $g_0: I \to \mathbf{R}$ lift $f \circ e_0 \circ (e|I)$, and let $h: [-1, 1] \to \mathbf{R}$ lift $f|[-1, 1]$. Since

$$e(h(-1)) = f(-1) = e(g_0(\tfrac{1}{2})),$$

we can choose the lift h so that $h(-1) = g_0(\frac{1}{2})$.

Now we construct lifts g_i of $f \circ e_i \circ (e|I)$ for $i = 1, 2$; these can be given explicitly by

$$g_1(t) = \begin{cases} g_0(t), & 0 \leqslant t \leqslant \frac{1}{2}, \\ h(\cos 2\pi t), & \frac{1}{2} \leqslant t \leqslant 1; \end{cases}$$

$$g_2(t) = \begin{cases} h(\cos 2\pi t), & 0 \leqslant t \leqslant \frac{1}{2}, \\ g_0(t), & \frac{1}{2} \leqslant t \leqslant 1. \end{cases}$$

Hence

$$\begin{aligned} \deg(f \circ e_0) &= g_0(1) - g_0(0) \\ &= g_0(1) - h(1) + h(1) - g_0(0) \\ &= g_2(1) - g_2(0) + g_1(1) - g_1(0) \\ &= \deg(f \circ e_2) + \deg(f \circ e_1). \blacksquare \end{aligned}$$

From this we deduce the following useful result.

13.2 Lemma *Let* $F: \Theta \to \mathbf{C}$ *be an embedding, let* $a, b \in \mathbf{C} - F(\Theta)$. *If two of* $F(S^1)$, $F(U)$ *and* $F(L)$ *fail to separate* a *from* b, *so does the third.*

Proof. Consider the map $f: \Theta \to S^1$ defined by

$$f(z) = N\left(\frac{F(z) - a}{F(z) - b}\right).$$

By Eilenberg's criterion (Theorem 9.4), $F(S^1)$ fails to separate a from b if and only if $f|S^1 = f \circ e_0$ is nullhomotopic, i.e. just when $\deg(f \circ e_0) = 0$; similarly for U and L. But by Lemma 13.1, if two of these degrees vanish, so does the third. ■

FIRST ALTERNATIVE PROOF (AFTER DIEUDONNÉ)

We have already noted that Eilenberg's criterion does not by itself show that for any Jordan curve $J \subset \mathbf{C}$, $\mathbf{C} - J$ has at most two components. In our first proof, this will be deduced from the following result, which is itself of some importance.

13.3 Proposition *If $J \subset \mathbf{C}$ is a Jordan curve, the frontier of any component of $\mathbf{C} - J$ is J.*

Proof. Let A be such a component. By Lemmas 9.1 and 9.2, A is open in $\mathbf{C}$ and $A \cup J$ is closed. Hence $\operatorname{Fr} A \subset J$.

Conversely, for any $x \in J$ and any neighborhood $U(x, \varepsilon)$, we can find an open arc α with $x \in \alpha$ and $\alpha \subset J \cap U(x, \varepsilon)$. By the corollary to Theorem 9.4, $\mathbf{C} - (J - \alpha)$ is path-connected. Hence there is a path in $\mathbf{C} - (J - \alpha)$ joining a point a of A to x. Since A is open, there will be some first point y on the path, not in A. Then $y \in \operatorname{Fr} A \subset J$, so $y \in \alpha$. Thus $\operatorname{Fr} A$ intersects $\alpha \subset U(x, \varepsilon)$; since this holds for all $\varepsilon > 0$, and $\operatorname{Fr} A$ is closed, by the corollary to Lemma 1.5, it follows that $x \in \operatorname{Fr} A$. This holds for all $x \in J$, so $J \subset \operatorname{Fr} A$. ■

We are now ready for our first alternative proof of the Jordan curve theorem.

Special case. *Let J be a Jordan curve in $\mathbf{C}$ which contains a straight line segment.* Let c lie on such an (open) segment γ'; let $r < d(c, J - \gamma')$. Describe a circle about c with radius r; denote by γ the diameter in which it meets γ', by α and β the semicircles on either side of γ, and by $\delta = J - \operatorname{Int}_J \gamma$ the rest of J (Fig. 13.2).

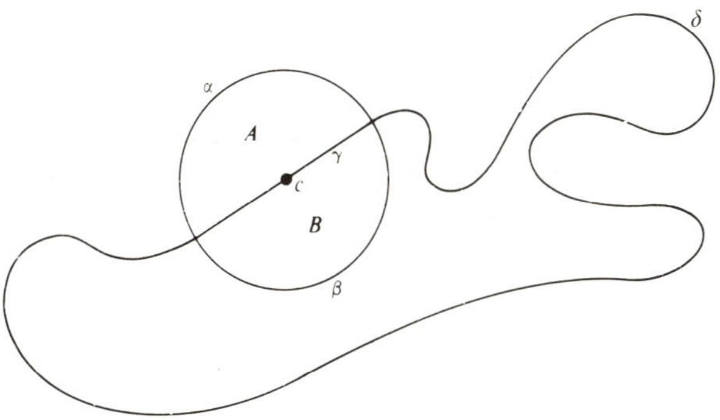

Fig. 13.2

Evidently $U(c, r) - \gamma$ has exactly two components A, B; we may suppose $\mathrm{Fr}\, A = \alpha \cup \gamma$, $\mathrm{Fr}\, B = \beta \cup \gamma$. By Proposition 13.3, every component of $\mathbf{C} - J$ meets $U(c, r)$, and so must contain A or B. Thus $\mathbf{C} - J$ has at most two components.

Choose $a \in A$, $b \in B$; it will suffice to show that $J = \gamma \cup \delta$ separates a from b. Now $\alpha \cup \delta$ does not meet the straight line segment ab, so does not separate the points, while $\alpha \cup \gamma$ clearly does separate them. Now by Lemma 13.2 it follows that $\gamma \cup \delta$ separates a from b. ■

General case. $J \subset \mathbf{C}$ *any Jordan curve.* Choose two distinct points of J, and draw the straight segment joining them. If this lies on J, we are in the special case above. If not, say $c \notin J$; since J is closed, there are a last point x on the segment before c and a first point y on the segment after c. Denote by γ the segment xy and by α, β the two (closed) arcs of J joining x to y (Fig. 13.3).

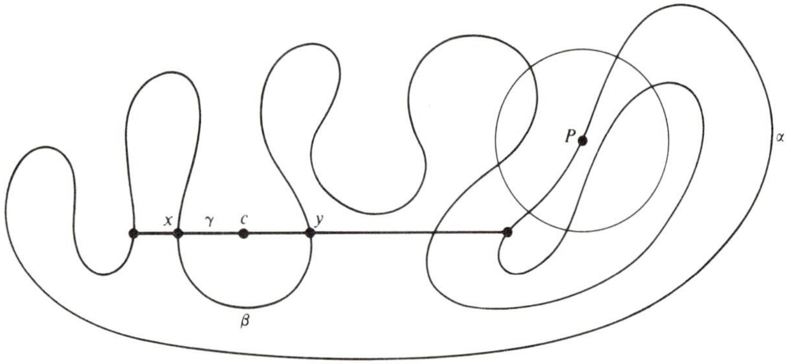

Fig. 13.3

Let $P \in \alpha$, and $r < d(P, \beta \cup \gamma)$. By Proposition 13.3, all components of $\mathbf{C} - J$ meets $U(P, r)$; the same holds with J replaced by $\alpha \cup \gamma$. But the theorem has already been established for the Jordan curve $\alpha \cup \gamma$. Thus it is enough to show that any two points a, b of $U(P, r) - J$ are separated by $\alpha \cup \beta$ if and only if they are separated by $\alpha \cup \gamma$. Since $\beta \cup \gamma$ does not meet the straight segment $ab \subset U(P, r)$, it does not separate a from b. The desired result now follows from Lemma 13.2. ■

POINT SETS IN $\mathbf{R}_n$ AND S_n

The idea of our second proof is to relate the question about our Jordan curve in the plane to one about its position in three-dimensional space, and then to solve the latter. The details involve some interrelation between spheres and euclidean spaces, as in Chapter 12.

Although the use of algebra in this book has been kept to a minimum until now, familiarity with quotient groups will be indispensable in the next chapter, and useful here. If A is an (abelian) group and B a subgroup, then one can define a "quotient group" A/B, and a quotient map $p: A \to A/B$. This is characterized up to isomorphism by either of the following two properties:

i) For any homomorphism $f: A \to G$ such that $f(B) = 0$, there is a (unique) homomorphism $\phi: A/B \to G$ with $f = \phi \circ p$. Compare the universal property of free abelian groups!

ii) The sequence

$$0 \to B \to A \xrightarrow{p} A/B \to 0$$

is exact.

The reader familiar with quotient groups will easily check these properties for himself. Now for any nonempty space X and point P, the unique map $c: X \to P$ induces an inclusion

$$\mathbf{Z} \cong H^0(P) \to H^0(X).$$

We will denote the corresponding quotient group by $\tilde{H}^0(X)$. In the next result we use an analogue of Theorem 8.1 for these groups, whose proof we leave as an exercise (Exercise 13.7).

Recall that S^n is the subset of $\mathbf{R}^{n+1}$ defined by

$$\sum_1^{n+1} x_i^2 = 1.$$

We identify $\mathbf{R}^{n+1}$ with the subset of $\mathbf{R}^{n+2}$ given by $x_{n+2} = 0$, and hence S^n with a subset of S^{n+1}.

13.4 Theorem *If K is a subset of S^n,*

$$\tilde{H}^0(S^n - K) \cong H^1(S^{n+1} - K).$$

Proof. Let D_+^{n+1}, D_-^{n+1} be the subspaces of S^{n+1} defined, respectively, by $x_{n+2} \geqslant 0$ and $x_{n+2} \leqslant 0$: we think of them as upper and lower hemispheres. Since D_+^{n+1} is closed in S^{n+1}, $D_+^{n+1} - K$ is closed in $S^{n+1} - K$. Hence by the Mayer-Vietoris theorem (Exercise 13.7), the sequence

$$\tilde{H}^0(D_+^{n+1} - K) \oplus \tilde{H}^0(D_-^{n+1} - K) \to \tilde{H}^0(S^n - K) \xrightarrow{\Delta} H^1(S^{n+1} - K) \to$$
$$\to H^1(D_+^{n+1} - K) \oplus H^1(D_-^{n+1} - K)$$

is exact. We wish to show that Δ is an isomorphism; this will follow by exactness if the outer terms are zero (by Lemma 2.1). Thus it will be enough to show that $D_+^{n+1} - K$ is contractible.

Orthogonal projection of $\mathbf{R}^{n+2}$ on $\mathbf{R}^{n+1}$ gives a homeomorphism of D_+^{n+1} on D^{n+1} and hence of $D_+^{n+1} - K$ on $D^{n+1} - K$. But $D^{n+1} - K$ is convex, and hence contractible. ∎

Note in particular that if $J \subset S^2$ is a Jordan curve,

$$\tilde{H}^0(S^2 - J) \cong H^1(S^3 - J).$$

As in Theorem 11.7, to show that $S^2 - J$ has two path components, we must show that these groups are infinite cyclic. This we will do using the following trick.

13.5 Theorem *Let $K \subset \mathbf{R}^m$, $L \subset \mathbf{R}^n$ be closed subsets, $f\colon K \to L$ a homeomorphism. Then there is a homeomorphism h of $\mathbf{R}^{m+n}$ onto itself such that for $k \in K$, $h(k, 0) = (0, f(k))$.*

Proof. Regard f as a map $K \to \mathbf{R}^n$. Since K is closed in $\mathbf{R}^m$, by applying Tietze's extension theorem (corollary to Theorem 7.4) to the components, we deduce that f has a continuous extension $g\colon \mathbf{R}^m \to \mathbf{R}^n$. We now define a homeomorphism h_1 of $\mathbf{R}^{m+n} = \mathbf{R}^m \times \mathbf{R}^n$ by

$$h_1(x, y) = (x, y + g(x)).$$

It is clear that this, and its inverse, are continuous. We have

$$h_1(k, 0) = (k, g(k)) = (k, f(k)).$$

Applying the same argument to f^{-1}, we obtain a homeomorphism h_2 of $\mathbf{R}^{m+n}$ with

$$h_2(0, f(k)) = (k, f(k)).$$

We now take $h = h_2^{-1} \circ h_1$ to prove the theorem. ∎

Two cases of this theorem are worthy of special comment. If K is compact, then the fact that K and $L = f(K)$ are closed follows from Lemma 1.9, without our needing to assume it. The case we will apply is, however, somewhat different.

Corollary *Let $f\colon \mathbf{R} \to \mathbf{R}^n$ be an embedding with closed image. Then there is a homeomorphism h of $\mathbf{R}^{n+1}$ with $h(0, \ldots, 0, t) = f(t)$ for $t \in \mathbf{R}$.* ∎

To apply this to our problem, we must first translate from euclidean spaces to spheres.

13.6 Lemma *Let N be the unit point on the x_{n+1}-axis; let $e\colon \mathbf{R}^n \to S^n - \{N\}$*

denote the homeomorphism obtained by stereographic projection from N. Let h be a homeomorphism of $\mathbf{R}^n$. *Define* $H: S^n \to S^n$ *by*

$$H(e(x)) = e(h(x)) \quad (x \in \mathbf{R}^n), \qquad H(N) = N.$$

Then H is a homeomorphism.

Proof. H is clearly bijective; if we can prove H continuous, the same argument will imply H^{-1} continuous, and hence that H is a homeomorphism.

Let U be open in S^n. If $N \notin U$, then $U = e(V)$, with V open in $\mathbf{R}^n$; if $N \in U$, then $S^n - U = e(K)$, with K compact in $\mathbf{R}^n$. Since h is a homeomorphism, $h^{-1}(V)$ is open (resp. $h^{-1}(K)$ is compact) in $\mathbf{R}^n$. Hence, in the first case, $H^{-1}(U) = e(h^{-1}(V))$ is open; in the second, $H^{-1}(S^n - U) = e(h^{-1}(K))$ is compact, hence closed, so again $H^{-1}(U)$ is open. Thus H is continuous. ■

SECOND ALTERNATIVE PROOF (AFTER DOYLE)

Let $J \subset S^2$ be a Jordan curve. We may assume (rotating S^2 if necessary) that J contains the north pole N. Then $J - \{N\}$ is homeomorphic to S^1 minus a point, and hence to $\mathbf{R}$, and is closed in $S^2 - \{N\}$ since J is closed in S^2. Thus there exists an embedding $f: \mathbf{R} \to \mathbf{R}^2$ with $J - \{N\} = e(f(\mathbf{R}))$, and hence with closed image.

By the corollary to Theorem 13.5, there is a homeomorphism h of $\mathbf{R}^3$, taking the x_3-axis to $f(\mathbf{R}) \subset \mathbf{R}^2 \subset \mathbf{R}^3$. Now by Lemma 13.6 this determines a homeomorphism H of S^3 taking the great circle C given by $x_1 = x_2 = 0$ to $J \subset S^2 \subset S^3$. Thus H induces a homeomorphism of $S^3 - J$ and $S^3 - C$. But by Theorem 12.4, $S^3 - C$ is homeomorphic to $S^1 \times \mathbf{C}$, and so, using Theorem 13.4, we have

$$\begin{aligned} \tilde{H}^0(S^2 - J) &\cong H^1(S^3 - J) \\ &\cong H^1(S^3 - C) \\ &\cong H^1(S^1 \times \mathbf{C}) \cong \mathbf{Z}, \end{aligned}$$

and $H^0(S^2 - J) \cong \mathbf{Z} \oplus \mathbf{Z}$. Since $S^2 \dot{-} J$ is l.p.c., $H^0(S^2 - J)$ is isomorphic by Theorem 4.7 to the group Map $(\pi_0(S^2 - J), \mathbf{Z})$. Hence $S^2 - J$ has just two components. ■

This last proof of the Jordan curve theorem leads to the question whether, given a Jordan curve J in $\mathbf{R}^2$, there is not a homeomorphism h of $\mathbf{R}^2$ on itself such that $h(J) = S^1$. The answer to this question is affirmative, but the proof is substantially harder than that of the Jordan curve theorem. The corresponding question for Jordan curves in $\mathbf{R}^3$ is answered negatively because of the existence of knots (though this, too, needs proof).

INVARIANCE OF (PLANE) DOMAINS

We now obtain some consequences of the Jordan curve theorem.

13.7 Lemma *Let $f: D^2 \to \mathbf{R}^2$ be injective and continuous. Then f is an embedding, and $f(\text{Int } D^2)$ is the bounded component of $\mathbf{R}^2 - f(S^1)$.*

Proof. Since D^2 is compact, it follows by Theorem 1.11 that f is an embedding. Hence $f(S^1)$ is a Jordan curve, and so separates $\mathbf{R}^2$ into two components. Also, Int D^2 is connected, hence so is $f(\text{Int } D^2)$; it does not meet $f(S^1)$ (f is injective), so it is contained in one component of $\mathbf{R}^2 - f(S^1)$.

If $f(\text{Int } D^2)$ is not the whole component of $\mathbf{R}^2 - f(S^1)$, then $\mathbf{R}^2 - f(D^2)$ contains points in both components of $\mathbf{R}^2 - f(S^1)$, so is not connected, contradicting the corollary to Theorem 9.4, since $f(D^2)$ is homeomorphic to D^2. As $f(D^2)$ is compact, and hence bounded, the result is now proved. ■

Corollary *With f as above, $f(\text{Int } D^2)$ is open in $\mathbf{R}^2$.* ■

We can now prove Brouwer's theorem on the "invariance of (plane) domains." The word *domain* here means a connected open subset (of $\mathbf{R}^2$).

13.8 Theorem *Let U be open in $\mathbf{R}^2$, and $h: U \to \mathbf{R}^2$ injective and continuous. Then h is an embedding and $h(U)$ is open in $\mathbf{R}^2$.*

Proof. Let $x \in U$; choose r so that $U(x, 2r) \subset U$—this is possible since U is open. Define $f: D^2 \to \mathbf{R}^2$ by

$$f(u + iv) = h(x_1 + ru, x_2 + rv).$$

Then f is continuous and injective since h is, so by the corollary to Lemma 13.7, $f(\text{Int } D^2)$ is open in $\mathbf{R}^2$. Hence $h(x) = f(0)$ is an interior point of $h(U)$. Since this is true for all $x \in U$, $h(U)$ is open in $\mathbf{R}^2$.

For h to be an embedding, we must show that V open in U implies $h(V)$ open in $h(U)$. But V open in U implies V open in $\mathbf{R}^2$; what we have already proved now implies $h(V)$ open in $\mathbf{R}^2$, and so in $h(U)$. ■

The result is, of course, trivial if h extends to a homeomorphism of $\mathbf{R}^2$ onto itself, since this would automatically take open sets to open sets. An example of h which does not extend is given by $U = U_1(0) \subset \mathbf{C}$, $h(z) = z/(1 - |z|)$. Part of the result can be paraphrased: if $V \subset \mathbf{R}^2$ is homeomorphic to an open subset of $\mathbf{R}^2$, then V is open. This emphasizes the homogeneity of $\mathbf{R}^2$. The corresponding property fails to hold, for example, for the space Θ introduced at the beginning of this chapter.

FURTHER DEVELOPMENTS

Since the Jordan curve theorem is really only a special case of the duality theorem, generalizations of the latter give generalizations of the former, too; also of Theorem 13.4. Similarly, the invariance of domain can be generalized straightforwardly to $\mathbf{R}^n$. More interesting is the result that for any Jordan curve J in $\mathbf{R}^2$, there is a homeomorphism h of $\mathbf{R}^2$ with $h(J) = S^1$, the "Schönflies" theorem. (See Newman for a proof. The first step is Exercise 13.3 below.) The same ideas prove the result also for arcs (homeomorphic to I). In $\mathbf{R}^3$, the result fails not only for S^1 (knots) but also for I (wild arcs) and for S^2 (wild spheres); see, for example, the classic paper of Fox and Artin. Conditions on an embedding $f: S^{n-1} \to \mathbf{R}^n$ which imply that there exists a homeomorphisn h of $\mathbf{R}^n$ with $h \circ f =$ identity are of great interest. See for example the paper of Brown.

Brown, M., "A Proof of the Generalized Schönflies Theorem," *Bull. Amer. Math. Soc.*, **66**, 74–76, (1960).

Fox, R. H. and E. Artin, "Some Wild Cells and Spheres in Three-dimensional Space," *Ann. of Math.*, **49**, 979–990, (1948).

Newman, M. H. A., *Elements of the Topology of Plane Sets of Points*, Cambridge University Press, 1939 (2nd edn. 1951).

EXERCISES AND PROBLEMS

1. a) Show that for any embedding $f: \Theta \to \mathbf{C}$, $\mathbf{C} - f(\Theta)$ has three components, whose respective frontiers are $f(S^1)$, $f(U)$, and $f(L)$.
   b) Let G be the graph with four vertices A, B, C, and D and six edges AB, AC, AD, BC, BD, and CD. Show that for any embedding $f: G \to \mathbf{C}$, $\mathbf{C} - f(G)$ has four components a, b, c, and d, where the frontier of a is the triangle BC, CD, DB, etc.
   c) Let H be the graph with five vertices A, B, C, D, and E and ten edges, one joining each pair of vertices. Show that there is no embedding $f: H \to \mathbf{C}$ by consideration of which component of $\mathbf{C} - f(G)$ would contain $f(E)$ (for example, if $f(E) \in a$, show that $f(AE)$ must cross some other edge).
2. Let $J \subset \mathbf{C}$ be a Jordan curve, A one of its complementary components. Show that given $x \in J$ and $\varepsilon > 0$, there exists $\delta > 0$ such that two points $p, q \in A$ with $d(x, p)$ and $d(x, q) < \delta$ can be joined by a path in $A \cap U(x, \varepsilon)$. [*Hint:* The question relates to whether $J \cup \{z: |z - x| = \varepsilon\}$ separates p from q. Choose sets to which Theorem 9.5 applies.]
3. Using the preceding exercise, construct a path $f: [0, 1] \to \mathbf{C}$ with $f([0, 1[) \subset A$ and $f(1) = x$.
4. Let $M^r \subset \mathbf{R}^n$ be a manifold, i.e. each point of M has a neighborhood homeomorphic to an open subset of $\mathbf{R}^r$. Show that each point of $M^r \times 0 \subset \mathbf{R}^{n+r}$ has a neighborhood U in $\mathbf{R}^{n+r}$ with a homeomorphism $\phi: U \to \mathbf{R}^{n+r}$ such that $\phi(U \cap M) = \mathbf{R}^r \times 0 \subset \mathbf{R}^{n+r}$.

5. Does invariance of domain (Theorem 13.8) continue to hold if $\mathbf{R}^2$ is replaced by $\mathbf{R}$? by I? by Int D^2? by D^2? by S^2?
6. A *surface* is a space each point x of which has a neighborhood N homeomorphic to D^2, say by $\phi: N \to D^2$. Show that if $\phi(x) \in S^1$ for one such "chart," then it is so for them all. [*Hint:* Use invariance of domain.] Such points x are called *boundary points* of the surface.
7. Prove that Theorem 8.1 continues to hold if H^0 is replaced by $\tilde{H}^0$ throughout.
8. Give examples to show that invariance of domain does not hold for θ-curves: i.e. give open sets $A \subset \theta$ and continuous injective $f: A \to \theta$ such that
   a) f is not an embedding
   b) $f(A)$ is not open in θ.

*9. Discuss the analogue of Proposition 13.3 for the "lakes of Wada".

10. Show that if K is compact in $\mathbf{R}$, and SK is defined (as in Exercise 8.3) as the union of segments joining points of K to $(0, 1)$ and to $(0, -1)$ in $\mathbf{R}^2$, then $\mathbf{R} - K \subset \mathbf{R}^2 - SK$ induces a bijection of π_0. Deduce that there is an isomorphism $\tilde{H}_0(\mathbf{R} - K) \to \tilde{H}^0(K)$, and hence that $\tilde{H}^0(K)$ is a free abelian group. Now try Exercise 4.7 again.

14
FURTHER DUALITY PROPERTIES

INTRODUCTION

As well as deducing the number of path components of $\mathbf{C} - K$ or $S^2 - K$ from properties of K, it is possible and interesting to relate connectedness of a compact set K to properties of $\mathbf{C} - K$. In this chapter we describe a new topological invariant $H_1(X)$ of a space X, and outline a second duality theorem, analogous to the one already given. This result is applied in the next chapter to discuss the formulation of Cauchy's theorem in complex analysis. We will not give proofs in this chapter, partly for reasons of space, and partly because it is no harder to discuss problems in any number of dimensions. The reader interested in a fuller account should consult one of the more advanced books listed at the end of the chapter.

THE GROUP $H_1(X)$

The duality theorem already given states that

$$D: \tilde{H}_0(\mathbf{C} - K) \to H^1(K)$$

is an isomorphism. Our idea here will be to obtain an isomorphism

$$I: H_1(\mathbf{C} - K) \to H^0(K),$$

where H_1 is yet to be discussed. To guide our definition, note that $H^0(X)$ and $H^1(X)$ are defined as groups of homotopy classes of maps $X \to \mathbf{Z}$ and $X \to S^1$ respectively; for $H_0(X)$ we had to use points of X (or, maps of a point into X), an equivalence relation, and some algebra. Analogy suggests that for $H_1(X)$ we should use maps of I or S^1 into X, an equivalence relation, and some algebra; it will be technically more convenient to use I. The equivalence relation will then have to permit gluing two intervals at a common endpoint; also homotopies fixing endpoints. The definition below is equivalent to this, but simpler.

For any space X, we have the set Map (I, X) of all continuous maps

$f: I \to X$. Define the group $C_1(X) = F(\text{Map}(I, X))$. Define ∂': Map $(I, X) \to F(X) = C_0(X)$ by

$$\partial'(f) = i(f(1)) - i(f(0)),$$

and let $d_1 : C_1(X) \to C_0(X)$ be the corresponding homomorphism.

We leave as an exercise to the reader the following reformulation of our definition of $H_0(X)$.

14.1 Lemma *The following sequence is exact:*

$$C_1(X) \xrightarrow{d_1} C_0(X) \xrightarrow{F(p)} H_0(X) \longrightarrow 0. \blacksquare$$

The definition of H_1 is modeled on this. Write T for the triangle $0 \leqslant y \leqslant x \leqslant 1$ in $\mathbf{R}^2$, and define homeomorphisms of I with the edges of T by

$$\partial_0(t) = (t, 0), \qquad \partial_1(t) = (t, t), \qquad \partial_2(t) = (1, t)$$

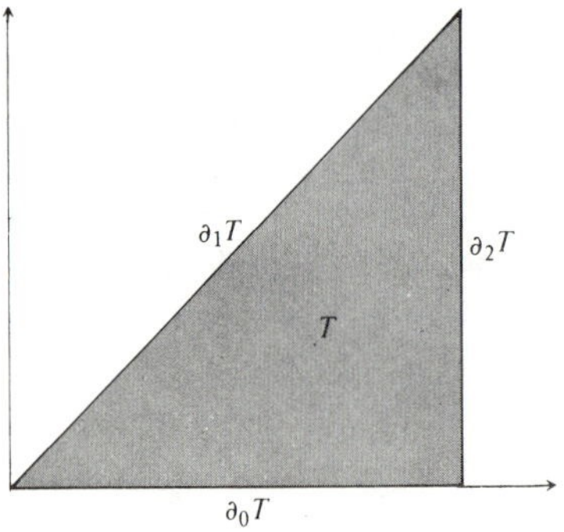

Fig. 14.1

(Fig. 14.1). Now write $C_2(X) = F(\text{Map}(T, X))$, and let $d_2 : C_2(X) \to C_1(X)$ be the homomorphism such that for $\phi: T \to X$ we have

$$d_2(i(\phi)) = i(\phi \circ \partial_0) - i(\phi \circ \partial_1) + i(\phi \circ \partial_2).$$

The reader should verify for himself the following simple result as a check in understanding the notation, and the reason for the position of the minus sign.

14.2 Lemma *We have* $d_1 \circ d_2 = 0$. $\blacksquare$

An element of Ker d_1 is called a 1-*cycle*; of Im d_2 a 1-*boundary*. Define $H_1(X)$ to be the quotient group

$$H_1(X) = \text{Ker } d_1/\text{Im } d_2;$$

by the lemma, Im $d_2 \subset$ Ker d_1.

The formal properties of H_1 can now be obtained like those of H_0. We give below the statements, and hints which should enable the studious reader to prove the results for himself.

14.3 Lemma *Any continuous map $f: X \to Y$ induces a homomorphism*

$$f_*: H_1(X) \to H_1(Y);$$

the identity map induces the identity homomorphism, and if $g: Y \to Z$, then $(g \circ f)_* = g_* \circ f_*$.

Note that f induces (by composition) a map

$$f_*: \mathrm{Map}\,(I, X) \to \mathrm{Map}\,(I, Y),$$

and thus a homomorphism

$$F(f_*): C_1(X) \to C_1(Y).$$

Similarly, there are induced homomorphisms of C_0, C_2, and you should check that the diagram below commutes. The assertions follow by algebraic manipulations.

$$\begin{array}{ccccc} C_2(X) & \to & C_1(X) & \to & C_0(X) \\ \downarrow & & \downarrow & & \downarrow \\ C_2(Y) & \to & C_1(Y) & \to & C_0(Y) \end{array} \blacksquare$$

14.4 Lemma *If $f_0 \simeq f_1: X \to Y$, then $f_{0*} = f_{1*}: H_1(X) \to H_1(Y)$.*

Indeed, let $H: X \times I \to Y$ be a homotopy. For each $P \in X$, $t \mapsto H(P, t)$ is a path in Y, giving an element of $C_1(Y)$. We thus obtain a homomorphism

$$s_0: C_0(X) \to C_1(Y).$$

Similarly, for each $f \in \mathrm{Map}\,(I, X)$, we define sf and $s'f \in \mathrm{Map}\,(T, Y)$ by

$$sf(u, v) = H(f(u), v),$$

$$s'f(u, v) = H(v, f(u)).$$

Taking the class of f to $i(sf) - i(s'f)$ gives a map

$$s_1: C_1(X) \to C_2(Y).$$

Now

$$d_2 s_1(x) = f_{1*}(x) - f_{0*}(x) - s_0 d_1(x),$$

and the result follows easily. ■

PROPERTIES OF $H_1(X)$

The following result is much harder, and we will not say much about the proof. See Exercises 14.2 and 14.3.

14.5 Mayer-Vietoris Theorem *Let X_1 and X_2 be open subsets of X with $X_1 \cup X_2 = X$, $X_1 \cap X_2 = Y$. Then there is an exact sequence (with maps as in Theorem* 8.1)

$$H_1(Y) \to H_1(X_1) \oplus H_1(X_2) \to H_1(X) \xrightarrow{\Delta} H_0(Y) \to H_0(X_1) \oplus H_0(X_2) \to H_0(X) \to 0.$$

Note that for Theorem 8.1, X_1 and X_2 were required to be closed in X; here we want them to be open. As before, the key part of the proof is the definition of the map Δ. The idea is, given maps of intervals (triangles) into X, to chop them up—or subdivide them—into smaller intervals (or triangles), each of which is mapped entirely into X_1 or X_2. Now if x is a chain in X representing $\xi \in H_1(X)$, we subdivide x to obtain $x_1 + x_2$ with $x_i \in C_1(X_i)$, and then $d_1x_1 + d_1x_2 = 0$, so that

$$d_1x_1 \in C_0(X_1 \cap X_2).$$

We now define $\Delta\xi$ as the class of d_1x_1. The whole proof consists of a careful use of this subdivision process. ■

The opposite of subdivision is also of interest, and enables us to relate the definitions above to ones using S^1 instead of I.

14.6 Lemma *Let X be path-connected. Any element of $H_1(X)$ is represented by a loop $f: I \to X$.*

For given two loops with $f_2(0) = f_1(1)$, form the composite $f_1 * f_2$ as in Chapter 5. It is easy to construct a triangle with

$$d_2(\phi) = i(f_1 * f_2) - i(f_1) - i(f_2).$$

Similarly $-i(f)$ can be replaced by $i(R \circ f)$. Now apply this to any 1-cycle: the terms can be collected into loops, and the loops can be joined using path connectedness of X. ■

However, it is not so simple to write down the equivalence relation on loops which defines $H_1(X)$. To compose loops, one must choose a base point; then modulo homotopy one can define the fundamental group $\pi_1(X)$, which is generally nonabelian. The most geometrical way to obtain $H_1(X)$ is to use bordism, and seek to obtain a map into X of an (oriented) surface, whose boundary loops are a given set of loops in X. For such developments, see the references at the end of the chapter. We now continue developing the analogy between H_1 and H^1.

14.7 Lemma *There is a natural homomorphism*

$$H^1(X) \to \operatorname{Hom}\big(H_1(X), \mathbf{Z}\big).$$

If X is l.p.c. this is injective.

First, one computes (e.g. using Theorem 14.5) $H_1(S^1) \cong \mathbf{Z}$. Then each $f: X \to S^1$ induces

$$f_*: H_1(X) \to H_1(S^1) \cong \mathbf{Z},$$

which depends only on the homotopy class of f, by Theorem 14.5, and is easily seen to be additive. If $f_* = 0$, then for each loop $l: S^1 \to X$ we have $0 = (f \circ l)_*$, from which it follows that $\deg(f \circ l) = 0$; by Theorem 7.1, f is nullhomotopic. ■

One might hope for an isomorphism under some condition like that needed for Theorem 4.8. This can indeed be obtained from the standard theory of covering spaces.

DUALITY

Now we construct our second duality map. Let K be a compact subset of $\mathbf{C}$, $z \in K$, $f: I \to \mathbf{C} - K$. We define a map $g_z: I \to S^1$ by

$$g_z(t) = N\big(z - f(t)\big).$$

This lifts to a continuous map $\tilde{g}_z: I \to \mathbf{R}$; we define the *index* of f with respect to z to be $\tilde{g}_z(1) - \tilde{g}_z(0)$. Clearly this does not depend on the choice of lift, and it depends continuously on z. Thus it gives an element of $\operatorname{Map}(K, \mathbf{R})$. Extending by the universal property, we obtain a homomorphism

$$I_1: C_1(\mathbf{C} - K) \to \operatorname{Map}(K, \mathbf{R}).$$

14.8 Lemma *We have $I_1 \circ d_2 = 0$. There is a commutative diagram*

$$\begin{array}{ccc} C_1(\mathbf{C} - K) & \xrightarrow{I_1} & \operatorname{Map}(K, \mathbf{R}) \\ \big\downarrow d_1 & & \big\downarrow e_* \\ C_0(\mathbf{C} - K) & \xrightarrow{D_3} & \operatorname{Map}(K, S^1) \end{array}$$ ■

Corollary If $d_1 z = 0$, $I_1 z \in \operatorname{Map}(K, \mathbf{Z})$. *Hence I_1 induces a map*

$$I_2: H_1(\mathbf{C} - K) \to \operatorname{Map}(K, \mathbf{Z}) = H^0(K).$$ ■

Then I_2 is the duality map. We can now state the second duality theorem, and show how it is related to the first.

14.9 Theorem *For any compact subset K of $\mathbf{C}$, $I\colon H_1(\mathbf{C} - K) \to H^0(K)$ is an isomorphism. Moreover, if K_1 and K_2 are both compact subsets of $\mathbf{C}$, the following diagram commutes:*

$$\begin{array}{ccccccccccc}
H_1(\mathbf{C}-K_1-K_2) & \to & H_1(\mathbf{C}-K_1)\oplus H_1(\mathbf{C}-K_2) & \to & H_1(\mathbf{C}-K_1\cap K_2) & \to & \tilde{H}_0(\mathbf{C}-K_1-K_2) & \to & \tilde{H}_0(\mathbf{C}-K_1)\oplus\tilde{H}_0(\mathbf{C}-K_2 & \to & \tilde{H}_0(\mathbf{C}-K_1\cap K_2). \\
\downarrow & & \downarrow & & \downarrow & & \downarrow & & \downarrow & & \downarrow \\
H^0(K_1\cup K_2) & \to & H^0(K_1)\oplus H^0(K_2) & \to & H^0(K_1\cap K_2) & \to & H^1(K_1\cup K_2) & \to & H^1(K_1)\oplus H^1(K_2) & \to & H^1(K_1\cap K_2).
\end{array}$$

where the two sequences are the Mayer-Vietoris sequences of Theorems 8.1 *and* 14.5 *and the vertical maps are duality isomorphisms.*

Note that the first map in the lower sequence is injective, and the last in the upper sequence is surjective, so we can add 0 to each end of each sequence and retain exactness, as was observed earlier in special cases. Commutativity of the middle square is proved by starting with a cycle in $\mathbf{C} - (K_1 \cap K_2)$ which is the sum of chains in $\mathbf{C} - K_1$ and $\mathbf{C} - K_2$, and applying the diagram of Lemma 14.8. ■

Corollary *If K is a closed subset of S^2, there is an isomorphism*

$$H_1(S^2 - K) \to \tilde{H}^0(K).$$

For if $K = S^2$ the result is clear; otherwise rotate until $\infty \notin K$. Since K is closed in S^2, it is compact, so Theorem 14.9 is applicable. We leave the rest of the proof as an exercise (Exercise 14.5). ■

PLANE DOMAINS

One application of the above is to the topology of plane domains, i.e. connected open subsets of $\mathbf{C}$. These are of interest in complex analysis (see also next chapter). Here we discuss the relation between the open set U, its closure, and its frontier. As in Chapter 12, the natural containing space in many ways is S^2. If the closure of U is the whole of S^2, then the complement of U in S^2 coincides with its frontier, and $H_1(U) \cong \tilde{H}^0(S^2 - U)$ by the above corollary. Otherwise, we can rotate S^2 so that ∞ is not in the closure. This amounts to having U a bounded subset of $\mathbf{C}$. Recall that $Cl(U)$ denotes the closure of U (the same in $\mathbf{C}$ or in S^2), and $\operatorname{Fr} U = Cl(U) - U$ its frontier.

14.10 Theorem *Let U be a bounded connected open subset of $\mathbf{C}$. Then there are isomorphisms*

$$\tilde{H}^0(\operatorname{Fr} U) \cong \tilde{H}^0(\mathbf{C} - U) \cong H_1(U),$$

and these groups vanish if and only if $H^1(U)$ does.

Proof. Since U is connected, so is $Cl(U)$ (cf. Exercise 3.9). The Mayer-Vietoris sequence of $(\mathbf{C}; \mathbf{C} - U, Cl(U))$ (with $\tilde{H}^0$; see Exercise 13.7),

$$\tilde{H}^0(\mathbf{C}) \to \tilde{H}^0(\mathbf{C} - Cl(U)) \oplus \tilde{H}^0(U) \to \tilde{H}^0(\operatorname{Fr} U) \to H^1(\mathbf{C}),$$

now gives the first isomorphism, and the corollary to Theorem 14.9 gives the second. By Lemma 14.7, if $H_1(U) = 0$, so is $H^1(U)$.

Now suppose conversely that $H^1(U) = 0$ and, if possible, that $\tilde{H}^0(\mathbf{C} - U) \neq 0$. Then $\mathbf{C} - U$ is disconnected. Let $\mathbf{C} - U = X \cup Y$ be a partition. If U is bounded by $|z| = R$, then Y, say, contains all points with $|z| \geqslant R$, so X is compact. By Theorem 1.12, $d(X, Y) > 0$, call it ε. Define

$$U(X) = \{z \in \mathbf{C} : d(z, X) < \varepsilon/3\},$$

$$U(Y) = \{z \in \mathbf{C} : d(z, Y) < \varepsilon/3\};$$

since $d(z, X)$ is continuous, $U(X)$ is open. Similarly, so is $U(Y)$. Let

$$K = \mathbf{C} - U(X) - U(Y).$$

This is closed and bounded (by R), hence compact. If $x \in X$ and $y \in Y$, then $x \in U(X)$, $y \in U(Y)$ lie in different components of $\mathbf{C} - K$, so are separated by K. By Eilenberg's criterion, if we define

$$f(z) = N\left(\frac{z - x}{z - y}\right),$$

then $f: K \to S^1$ is not homotopic to a constant. But since f is defined on U (x, y are not in U), and $H^1(U) = 0$, there is a nullhomotopy of $f: U \to S^1$, and this restricts to give a nullhomotopy on K. Thus we have a contradiction. ∎

When the groups vanish, the domain U is called simply connected. It can be shown that this is equivalent to the condition that all maps $S^1 \to U$ are nullhomotopic; also to the condition that for every Jordan curve $J \subset U$, U contains one complementary domain of J. Another equivalent (but deeper) condition is that U is homeomorphic to $\mathbf{C}$.

Corollary *For U a connected open subset of S^2, each component of $S^2 - U$ is simply-connected.*

For $H_1(S^2 - \bar{U}) = \tilde{H}^0(Cl(U)) = 0$. ∎

FURTHER DEVELOPMENTS

This chapter is a fairly straightforward introduction to traditional algebraic topology; we have deliberately used standard notations so that the reader will not be confused if he goes further. Suitable books are Lefschetz (somewhat old-fashioned, but very geometric), Maunder (perhaps the most suitable as sequel to this book). Hocking and Young, and Spanier (more thorough than the preceding, but rather much so for an introduction). We also mentioned bordism. For a general introduction to differential topology see Milnor; for bordism as such, try the first chapter of Conner and Floyd.

Conner, P. E., and E. E. Floyd, *Differentiable Periodic Maps*, Springer, Berlin, 1964.

Hocking, J. G. and G. S. Young, *Topology*, Addison-Wesley, Reading, Mass., 1961.

Lefschetz, Solomon, *Introduction to Topology*, Princeton University Press, 1949.

Maunder, C. R. F., *Algebraic Topology*, Van Nostrand, Princeton, 1970.

Milnor, J. W., *Topology from the Differentiable Viewpoint*, University Press of Virginia, 1965.

Spanier, E. H., *Algebraic Topology*, McGraw-Hill, New York, 1966.

EXERCISES AND PROBLEMS

1. Complete the proofs of Lemmas 14.1 through 14.4.
2. For any subsets X_1, X_2 of X with intersection Y, define

$$C_i^{\Delta}(X) = C_i(X_1) + C_i(X_2) \subset C_i(X) \qquad \text{for} \quad i = 1, 2.$$

Show (using properties of exact sequences) that if $H_1^{\Delta}(X)$, $H_0^{\Delta}(X)$ are defined using the C_i^{Δ}, then there is an exact sequence

$$H_1(Y) \to H_1(X_1) \oplus H_1(X_2) \to H_1^{\Delta}(X) \to H_0(Y) \to H_0(X_1) \oplus H_0(X_2) \to H_0^{\Delta}(X) \to 0.$$

3. Show that there are maps $H_1^{\Delta}(X) \to H_1(X)$, $H_0^{\Delta}(X) \to H_0(X)$, with the latter surjective. It is injective if and only if for every path in X with endpoints $P \in X_1$, $Q \in X_2$, there is a sequence of points

$$P = R_0, R_1, \ldots, R_{2k} = Q$$

such that R_{2i} can be joined to R_{2i+1} $(0 \leqslant i < k)$ by a path in X_1, and R_{2i-1} to R_{2i} $(1 \leqslant i \leqslant k)$ by a path in X_2. Show that this holds if X_1 and X_2 are both open in X. Need it hold if both are closed?

If for any $f\colon I \to X$ with $f(0) \in X_1$, $f(1) \in X_2$, we can subdivide I by

$$0 = a_0 < a_1 < \cdots < a_{2k} = 1$$

so that $f[a_{2i}, a_{2i+1}] \subset X_1$ and $f[a_{2i-1}, a_{2i}] \subset X_2$, show that $H_1^{\Delta}(X) \to H_1(X)$ is surjective.

4. Give a proof of Lemma 14.8.

5. Let K be a compact subset of $U_{\mathbf{C}}(0, R)$. The map $f: I \to \mathbf{C} - K$ given by $f(t) = Re(t)$ defines an element x of $H_1(\mathbf{C} - K)$. Show that $I_2(x)$ is the constant map $c: K \to \mathbf{Z}$ with $c(K) = 1$. Hence obtain an isomorphism of exact sequences

$$\begin{array}{ccccccccc} 0 \to & \mathbf{Z} & \to & H_1(\mathbf{C} - K) & \to & H_1(S^2 - K) & \to 0 \\ & \downarrow & & \downarrow I_2 & & \downarrow & \\ 0 \to & \mathbf{Z} & \to & H^0(K) & \to & \tilde{H}^0(K) & \to 0 \end{array}$$

[*Hint:* Use Mayer-Vietoris with $X_1 = \{z \in \mathbf{C}: |z| \leqslant R, z \notin K\}$ and $X_2 = \{z \in S^2: |z| \geqslant R\}$.]

6. Let $\Delta_n = \{x \in \mathbf{R}^{n+1}: x_i \geqslant 0, \Sigma\, x_i = 1\}$. Define $\partial_i: \Delta_{n-1} \to \Delta_n$ by

$$\partial_i(x_0, \ldots, x_{n-1}) = (x_0, \ldots, x_{i-1}, 0, x_i, \ldots, x_{n-1}) \qquad (0 \leqslant i \leqslant n).$$

Write $C_n(X) = F(\text{Map}\,(\Delta_n, X))$, and let $d_n: C_n(X) \to C_{n-1}(X)$ be the homomorphism such that for $\phi: \Delta_n \to X$,

$$d_n i(\phi) = \sum_{r=0}^{n} (-1)^r i(\phi \circ \partial_r).$$

Verify the analogues of Lemmas 14.1 through 14.4.

7. Let Γ be a graph with vertices $A_1, A_2, \ldots, A_n$ and straight edges. Let $M \subset \text{Map}\,(I, X)$ be the set of linear maps of I onto the edges A_iA_j with $i < j$; $C_1'(\Gamma) = F(M)$. Show (by induction, using the Mayer-Vietoris theorem) that $\text{Ker}\,(d_1|C_1'(\Gamma))$ maps isomorphically onto $H_1(\Gamma)$. (Compare with Exercises 8.25 through 8.27.)

8. Show that if U is open in $\mathbf{R}^n$, one obtains the same groups $H_1(U)$ by using, instead of all continuous maps $I \to U$, only those with continuous first-order partial derivatives. (Use the method of Exercise 3.12.) Try to show that you can restrict maps $T \to U$ too. (What do you need in order to do this?)

9. Describe the effect on H_1 of attaching an n-cell (compare with Exercises 8.9 and 8.10).

10. Use the preceding exercise to compute $H_1(M_g)$ and $H_1(N_h)$, where M_g, N_h are as in Exercise 8.12.

*11. Compute $H_1(P_n(\mathbf{R}))$ and $H_1(P_n(\mathbf{C}))$ using the two preceding exercises. [*Hint:* $P_n(\mathbf{R})$ is obtained from $P_{n-1}(\mathbf{R})$ by attaching an n-cell; $P_n(\mathbf{C})$ from $P_{n-1}(\mathbf{C})$ by adding a $2n$-cell.]

12. Use the middle square of the diagram of Theorem 14.9 to prove the following (Alexander's lemma). Let K, L be compact subsets of $\mathbf{C}$ intersecting in just two points P, Q; x, $y \in \mathbf{C} - K - L$ two points separated neither by K nor by L. Join by paths α in $\mathbf{C} - K$, β in $\mathbf{C} - L$. Then if the index of $\alpha * \beta$ with respect to P is 0, $K \cup L$ does not separate x from y.

13. Show that if Γ is a graph, then $H_1(\Gamma)$ is generated by the classes of loops which are Jordan curves.

14. For each of the following graphs, give an isomorphism of $H_1(\Gamma)$ on $\mathbf{Z}^r$ (for some r), and state which elements are represented by Jordan curves:

a) 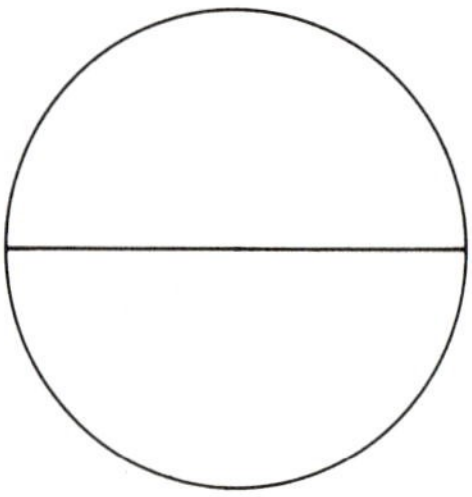

b) 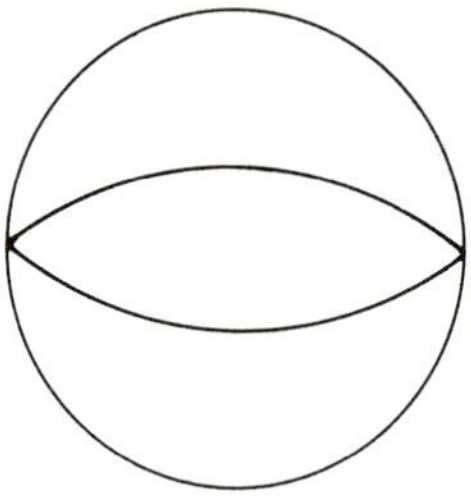

c) 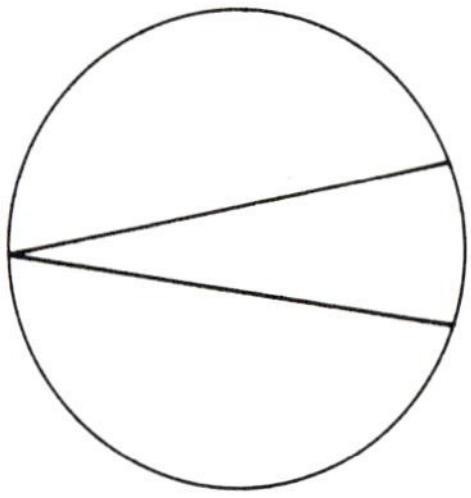

d) 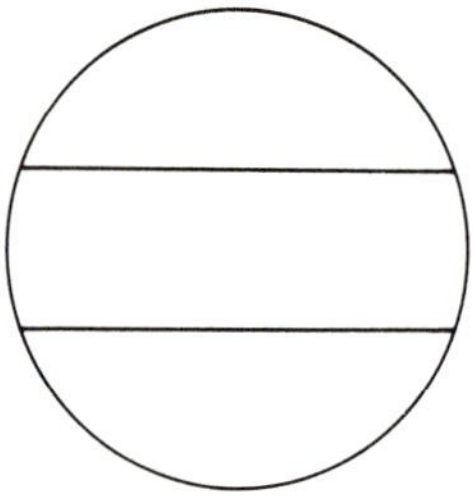

e) 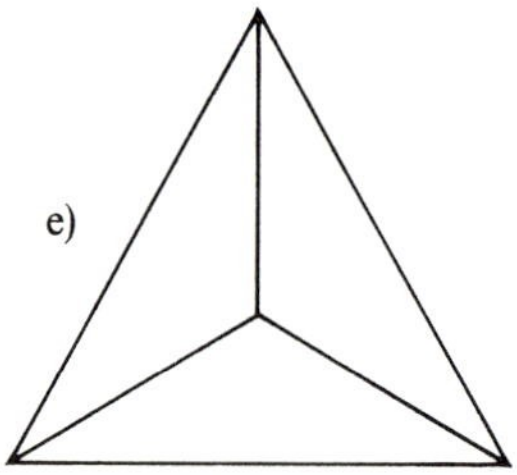

f) The graph of Exercise 11.9

g) The graph with five vertices, and one edge joining each pair

15
GEOMETRIC INTEGRATION THEORY

INTRODUCTION

The relation between the results of Chapter 14 and integrals comes from regarding a map $f: I \to \mathbf{C}$ as a possible path of integration. This requires of course that f be more than merely continuous. We will assume for simplicity that f is differentiable, though it is known that one can manage with the hypothesis that the path is rectifiable (i.e. has a finite length). Since several extra points arise in the complex-variable case, we begin this chapter by studying the case of two real variables.

LINE INTEGRALS IN $\mathbf{R}^2$

A line integral is usually written as

$$\int_s a\,dx + b\,dy,$$

where s denotes the path. It is convenient to take the path in parametrized form $x = x(t)$, $y = y(t)$ for (we may suppose) $0 \leqslant t \leqslant 1$; this amounts to saying we have a map $f: I \to \mathbf{R}^2$. The integral is then defined to be

$$\int_0^1 \left\{a\big(x(t), y(t)\big)\frac{dx(t)}{dt} + b\big(x(t), y(t)\big)\frac{dy(t)}{dt}\right\}dt,$$

and for this to make sense, the functions a and b must be defined on $f(I)$ and the derivatives dx/dt and dy/dt must exist.

To deal with the first point, we choose an open set $U \subset \mathbf{R}^2$, and will consider only maps $f: I \to U$ and functions a, b defined on U. For the second, we require of the map f that the derivatives dx/dt and dy/dt exist and are continuous on $[0, 1]$. Since the condition is imposed only on first-order derivatives, the set of such maps is commonly denoted by $C^1(I, U)$ but to avoid confusion with other notations in this book, we will write instead $\mathrm{Map}^d(I, U)$, where d stands for differentiable.

We will require conditions on the functions a and b also. It is enough at this stage to suppose that the first-order partial derivatives of a and b exist and are continuous on U. Expressions $a\,dx + b\,dy$ of this kind are called *differentiable* 1-*forms* on U; the vector space (over **R**) of all such forms is denoted by $\mathscr{C}^1(U)$.

GREEN'S THEOREM

The important theorem about line integrals is Green's theorem. This states, roughly speaking, that

$$\int_s a\,dx + b\,dy = \iint_A \left(\frac{\partial a}{\partial y} - \frac{\partial b}{\partial x}\right) dx\,dy,$$

where s denotes a loop, and A the region inside. It has been one of our main objectives to make the notion "inside" rigorous. Now let us study this theorem.

15.1 Lemma *Suppose $T \subset U$. Then*

$$\left(\int_{\partial_0} - \int_{\partial_1} + \int_{\partial_2}\right)(a\,dx + b\,dy) = \iint_T \left(-\frac{\partial a}{\partial y} + \frac{\partial b}{\partial x}\right) dx\,dy.$$

Proof. Clearly we can consider the terms in a and in b separately; the argument is the same in both cases, so it is enough to consider a. Now

$$\int_{\partial_0} a\,dx = \int_0^1 a(t, 0)\,dt,$$

$$\int_{\partial_1} a\,dx = \int_0^1 a(t, t)\,dt,$$

$$\int_{\partial_2} a\,dx = 0 \qquad (x \text{ is constant on } \partial_2).$$

Also,

$$a(t, t) - a(t, 0) = \int_0^t \frac{\partial a}{\partial y}(t, y)\,dy.$$

Hence

$$\left(\int_{\partial_0} - \int_{\partial_1} + \int_{\partial_2}\right) a\,dx = -\int_0^1 \left(\int_0^x \frac{\partial a}{\partial y}(x, y)\,dy\right) dx$$
$$= -\iint_T \frac{\partial a}{\partial y}\,dx\,dy,$$

as asserted. ■

This is a somewhat special case, but we can easily use it to obtain more. Let $\phi \in \mathrm{Map}^d(T, U)$. This means for now that both partial derivatives of both coordinates of ϕ exist and are continuous. To avoid confusion, use (u, v) for coordinates in T and (x, y) for coordinates in U. Then

$$\frac{\partial(x \circ \phi)}{\partial u}$$

is one of the derivatives in question.

Let

$$\omega = a(x, y)\,dx + b(x, y)\,dy$$

be a 1-form on U. If $f : I \to T \in \mathrm{Map}^d(I, T)$, then

$$\phi \circ f \in \mathrm{Map}^d(I, U),$$

and we have the line integral

$$\int_{\phi \circ f} \omega.$$

This can be expressed as a line integral along f as follows:

$$\int_{\phi\circ f} \omega = \int_0^1 \left\{a(\phi(f(t)))\frac{dx}{dt} + b(\phi(f(t)))\frac{dy}{dt}\right\} dt.$$

Now substitute

$$\frac{dx}{dt} = \frac{\partial x}{\partial u}\frac{du}{dt} + \frac{\partial x}{\partial v}\frac{dv}{dt},$$

$$\frac{dy}{dt} = \frac{\partial y}{\partial u}\frac{du}{dt} + \frac{\partial y}{dv}\frac{dv}{dt},$$

and we have a line integral $\int_f \phi^*\omega$, where

$$\phi^*\omega = \left(a(\phi(u, v))\frac{\partial x}{\partial u} + b(\phi(u, v))\frac{\partial y}{\partial u}\right)du + \left(a(\phi(u, v))\frac{\partial x}{\partial v} + b(\phi(u, v))\frac{\partial y}{\partial v}\right)dv.$$

Now apply Lemma 15.1 to $\phi^*\omega$. The left-hand side is

$$\left(\int_{\partial_0} - \int_{\partial_1} + \int_{\partial_2}\right)(\phi^*\omega) = \left(\int_{\phi\circ\partial_0} - \int_{\phi\circ\partial_1} + \int_{\phi\circ\partial_2}\right)(\omega),$$

by the above. The right-hand side is the double integral over T of $du\,dv$ multiplied by

$$\frac{\partial}{\partial u}\left\{a(\phi(u,v))\frac{\partial x}{\partial v} + b(\phi(u,v))\frac{\partial y}{\partial v}\right\} - \frac{\partial}{\partial v}\left\{a(\phi(u,v))\frac{\partial x}{\partial u} + b(\phi(u,v))\frac{\partial y}{\partial u}\right\}$$

$$= \frac{\partial a}{\partial u}\frac{\partial x}{\partial v} + a\frac{\partial^2 x}{\partial u\,\partial v} + \frac{\partial b}{\partial u}\frac{\partial y}{\partial v} + b\frac{\partial^2 y}{\partial u\,\partial v}$$

$$- \frac{\partial a}{\partial v}\frac{\partial x}{\partial u} - a\frac{\partial^2 x}{\partial v\,\partial u} - \frac{\partial b}{\partial v}\frac{\partial y}{\partial u} - b\frac{\partial^2 y}{\partial v\,\partial u}$$

$$= \left(\frac{\partial a}{\partial x}\frac{\partial x}{\partial u} + \frac{\partial a}{\partial y}\frac{\partial y}{\partial u}\right)\frac{\partial x}{\partial v} + \left(\frac{\partial b}{\partial x}\frac{\partial x}{\partial u} + \frac{\partial b}{\partial y}\frac{\partial y}{\partial u}\right)\frac{\partial y}{\partial v}$$

$$- \left(\frac{\partial a}{\partial x}\frac{\partial x}{\partial v} + \frac{\partial a}{\partial y}\frac{\partial y}{\partial v}\right)\frac{\partial x}{\partial u} - \left(\frac{\partial b}{\partial x}\frac{\partial x}{\partial v} + \frac{\partial b}{\partial y}\frac{\partial y}{\partial v}\right)\frac{\partial y}{\partial u}$$

$$= \left(\frac{\partial b}{\partial x} - \frac{\partial a}{\partial y}\right)\left(\frac{\partial x}{\partial u}\frac{\partial y}{\partial v} - \frac{\partial x}{\partial v}\frac{\partial y}{\partial u}\right).$$

Now if ϕ is injective, and the Jacobian positive, the usual rule for change of variable in a double integral shows that this double integral equals

$$\iint_{\phi(T)}\left(\frac{\partial b}{\partial x} - \frac{\partial a}{\partial y}\right)dx\,dy.$$

For any (twice differentiable) ϕ, we will *define* the double integral over ϕ of any function $c(x, y)$ to be

$$\iint_\phi c(x,y)\,dx\,dy = \iint_T c(\phi(u,v))\left(\frac{\partial x}{\partial u}\frac{\partial y}{\partial v} - \frac{\partial x}{\partial v}\frac{\partial y}{\partial u}\right)du\,dv.$$

Thus our integral can now be written

$$\iint_\phi\left(\frac{\partial b}{\partial x} - \frac{\partial a}{\partial y}\right)dx\,dy.$$

We further simplify by saying that for $\omega = a\,dx + b\,dy$ we define

$$\delta\omega = \left(\frac{\partial b}{\partial x} - \frac{\partial a}{\partial y}\right)dx\,dy.$$

We have thus shown that with this notation, Lemma 15.1 implies

15.2 Proposition *If U is open in $\mathbf{R}^2$, $\phi: T \to U$ has continuous second-order partial derivatives, and $\omega \in \mathscr{C}^1(U)$, then*

$$\int_{\phi \circ \partial_0} \omega - \int_{\phi \circ \partial_1} \omega + \int_{\phi \circ \partial_2} \omega = \iint_{\phi} \delta\omega. \qquad \blacksquare$$

This result suggests another improvement in notation. The integral $\int_s \omega$ is defined for $s \in \mathrm{Map}^1(I, U)$ and $\omega \in \mathscr{C}^1(U)$. We now extend it to $F(\mathrm{Map}^1(I, U))$ as a homomorphism by the universal property. Thus, if $\xi = \sum n_r i(x_r)$, $n_r \in \mathbf{Z}$, $x_r \in \mathrm{Map}^1(I, U)$, we have

$$\int_{\xi} \omega = \sum n_r \int_{x_r} \omega.$$

Similarly, $\iint_A \eta$ is defined as above for $A \in \mathrm{Map}^1(T, U)$ and η of the form

$$\eta = e(x, y)\, dx\, dy.$$

We extend $\iint_A \eta$ as a homomorphism to $F(\mathrm{Map}^1(T, U))$. Then Proposition 15.2 can be written as

$$\int_{\partial\phi} \omega = \iint_{\phi} \delta\omega$$

and is valid under the appropriate conditions of differentiability. Note that for $\phi^*\omega$ to be differentiable, we must suppose that ϕ has continuous second-order partial derivatives.

REFORMULATION IN TERMS OF HOMOLOGY

Let us now take stock. We have restricted the sequence

$$C_2(U) \xrightarrow{d} C_1(U) \xrightarrow{d} C_0(U)$$

used in Chapter 14 to define $H_1(U)$ and $H_0(U)$ by allowing only maps with continuous first-order partial derivatives. It can be shown that this yields the same homology groups as before (see Exercise 14.8). For the last result we had to insist on continuous second-order partial derivatives too. To avoid the necessity of continually stating our exact requirements, let us suppose from now on that our maps have continuous partial derivatives of all orders. The usual notations here are $C^\infty(I, U)$, $C^\infty(T, U)$ but we will simply write $\mathrm{Map}^d(I, U)$, $\mathrm{Map}^d(T, U)$ again, and $C_1^d(U)$, $C_2^d(U)$ for the corresponding free abelian groups. Again we obtain the same groups $H_0(U)$ and $H_1(U)$.

Similar comments apply to the 1-forms. We end up by writing $\mathscr{C}^0(U)$ for the vector space of real-valued functions on U with continuous partial derivatives of all orders; $\mathscr{C}^1(U)$ for the vector space of 1-forms

$$\omega = a\,dx + b\,dy$$

with $a \cdot b \in \mathscr{C}^0(U)$; and finally $\mathscr{C}^2(U)$ for expressions

$$p\,dx\,dy$$

with $p \in \mathscr{C}^0(U)$. We define linear maps

$$\mathscr{C}^0(U) \xrightarrow{\delta} \mathscr{C}^1(U) \xrightarrow{\delta} \mathscr{C}^2(U)$$

as follows. For $f \in \mathscr{C}^0(U)$,

$$\delta f = \frac{\partial f}{\partial x}dx + \frac{\partial f}{\partial y}dy.$$

For $\omega = a\,dx + b\,dy$ in $\mathscr{C}^1(U)$,

$$\delta\omega = \left(\frac{\partial b}{\partial x} - \frac{\partial a}{\partial y}\right)dx\,dy.$$

One reason for assuming infinite differentiability can now be seen, since if ω is r times continuously differentiable, we can only expect $\delta\omega$ to be $(r-1)$ times so. Observe that for any $f \in \mathscr{C}^0(U)$, $\delta\delta f = 0$ (compare with Lemma 14.1).

Now define

$$\mathscr{H}^0(U) = \text{Ker}\,(\delta\colon \mathscr{C}^0(U) \to \mathscr{C}^1(U))$$

$$\mathscr{H}^1(U) = \frac{\text{Ker}\,(\delta\colon \mathscr{C}^1(U) \to \mathscr{C}^2(U))}{\text{Im}\,(\delta\colon \mathscr{C}^0(U) \to \mathscr{C}^1(U))}$$

The first of these is easily described. If $\delta f = 0$, then

$$\frac{\partial f}{\partial x} = \frac{\partial f}{\partial y} = 0$$

on U. By the increment formula, we see that for any $P \in U$, f is constant near P. It follows that f is constant on each component of U and so determines a map $\pi_0(U) \to \mathbf{R}$. We deduce

$$\mathscr{H}^0(U) \cong \text{Map}\,(\pi_0(U), \mathbf{R}) \cong \text{Hom}\,(H_0(U), \mathbf{R}).$$

More interesting is the other group, which is related to the work we have just done.

15.3 Theorem *Integration defines an injective map*

$$J\colon \mathscr{H}^1(U) \to \text{Hom}\,(H_1(U), \mathbf{R}).$$

Remark. In fact, J is an isomorphism. This follows from de Rham's theorem, which is described in the references at the end of the chapter.

Proof. Let $\omega \in \mathscr{C}^1(U)$, with $\delta\omega = 0$. We know that for any $\xi \in C_1^d(U)$ we have the integral $\int_\xi \omega$. We are only interested in Ker d, that is, we suppose $\partial\xi = 0$. Now if $\xi = d\phi$, then

$$\int_\xi \omega = \int_{d\phi} \omega = \int_\phi \delta\omega = 0,$$

since $\delta\omega = 0$. Hence $\xi \to \int_\xi \omega$ induces a homomorphism

$$J(\omega) \colon H_1(U) \to \mathbf{R}.$$

Clearly, $J(\omega_1 + \omega_2) = J(\omega_1) + J(\omega_2)$. Now to show that J defines a homomorphism of $\mathscr{H}^1(U)$, it remains to check that if $\omega = \delta f$, then $J(\omega) = 0$. Now for any $\xi \in \mathrm{Map}^d(I, U)$,

$$\begin{aligned}\int_\xi \omega &= \int_\xi \frac{\partial f}{\partial x}dx + \frac{\partial f}{\partial y}dy \\ &= \int_0^1 \left(\frac{\partial f}{\partial x}\frac{dx}{dt} + \frac{\partial f}{\partial y}\frac{dy}{dt}\right)dt \\ &= \int_0^1 \frac{d}{dt} f\big(x(t), y(t)\big)\,dt \\ &= f\big(x(1), y(1)\big) - f\big(x(0), y(0)\big) \\ &= \int_{\partial\xi} f\end{aligned}$$

in an obvious notation. By linearity, this holds for any $\xi \in C_1^d(U)$. Hence if $\partial\xi = 0$, we have $\int_\xi \omega = 0$, as asserted.

Note the similarity with Lemma 14.8 of this argument. It remains to show that J is injective. Suppose ω such that $J(\omega) = 0$. We may suppose U connected; if it is not, we can apply the following argument separately to each component. Choose a point $P \in U$. For each $Q \in U$, there is (by Exercise 3.11) a path $\xi_Q \in \mathrm{Map}^d(I, U)$ from P to Q. We define

$$f(Q) = \int_{\xi_Q} \omega.$$

Now for any $\xi \in \mathrm{Map}^d(I, U)$, write $Q_0 = \xi(0)$ and $Q_1 = \xi(1)$. Then

$$i(\xi_{Q_0}) + i(\xi) - i(\xi_{Q_1}) = \eta \in C_1^d(U)$$

satisfies $d\eta = 0$. Since $J(\omega) = 0$, it follows that

$$0 = \int_\eta \omega = \int_{\xi_{Q_0}} \omega + \int_\xi \omega - \int_{\xi_{Q_1}} \omega$$

and so

$$\int_\xi \omega = f(Q_1) - f(Q_0).$$

It follows, in particular, that the definition of $f(Q)$ does not depend on the choice of path ξ_Q.

Take the path ξ to be horizontal. We then find that (provided the line segment $p \leqslant x \leqslant p + h$, $y = q$ lies in U), if we write $\omega = a\,dx + b\,dy$,

$$f(p + h, q) - f(p, q) = \int_0^1 a(p + th, q)h\,d\ .$$

Dividing by h, and letting $h \to 0$, we see that $\partial f/\partial x$ exists and equals $a(x, y)$. Similarly, $\partial f/\partial y$ exists and equals $b(x, y)$. Hence f has continuous partial derivatives of all orders (since a and b do). ■

THE THREE-DIMENSIONAL CASE

We will now discuss very briefly the corresponding situation in three dimensions. Here we can extend the above (just as Exercise 14.6 shows how to extend Lemma 14.1) to obtain, for U open in $\mathbf{R}^3$, maps

$$C_3^s(U) \xrightarrow{d} C_2^s(U) \xrightarrow{d} C_1^s(U) \xrightarrow{d} C_0^s(U) \to 0$$

with $d^2 = 0$; taking Ker d/Im d gives us groups $H_2(U)$, $H_1(U)$, and $H_0(U)$.

Similarly, we have

$$0 \to \mathscr{C}^0(U) \xrightarrow{\delta} \mathscr{C}^1(U) \xrightarrow{\delta} \mathscr{C}^2(U) \xrightarrow{\delta} \mathscr{C}^3(U)$$

where $\mathscr{C}^0(U)$ is the vector space of all real-valued functions f on U with continuous partial derivates of all orders, and $\mathscr{C}^1(U)$, $\mathscr{C}^2(U)$, and $\mathscr{C}^3(U)$ consist respectively of expressions

$$a_1\,dx_1 + a_2\,dx_2 + a_3\,dx_3,$$
$$b_1\,dx_2\,dx_3 + b_2\,dx_3\,dx_1 + b_3\,dx_1\,dx_2,$$
$$c\,dx_1\,dx_2\,dx_3,$$

with a_i, b_i, $c \in \mathscr{C}^0(U)$. An element of $\mathscr{C}^i(U)$ is called a *smooth i-form* on U. The maps δ are defined by

$$\delta f = \frac{\partial f}{\partial x_1}dx_1 + \frac{\partial f}{\partial x_2}dx_2 + \frac{\partial f}{\partial x_3}dx_3,$$

$$\delta(a_1\,dx_1 + a_2\,dx_2 + a_3\,dx_3) = \left(\frac{\partial a_3}{\partial x_2} - \frac{\partial a_2}{\partial x_3}\right)dx_2\,dx_3 + \left(\frac{\partial a_1}{\partial x_3} - \frac{\partial a_3}{\partial x_1}\right)dx_3\,dx_1 + \left(\frac{\partial a_2}{\partial x_1} - \frac{\partial a_1}{\partial x_2}\right)dx_1\,dx_2,$$

$$\delta(b_1\,dx_2\,dx_3 + b_2\,dx_3\,dx_1 + b_3\,dx_1\,dx_2) = \left(\frac{\partial b_1}{\partial x_1} + \frac{\partial b_2}{\partial x_2} + \frac{\partial b_3}{\partial x_3}\right)dx_1\,dx_2\,dx_3.$$

Observe the intimate relation with the familiar differential operators grad, curl, div.

Now we have the trivial formula (cf. discussion above): for $f \in \mathscr{C}^0(U)$, $\xi \in C_1^s(U)$,

$$\int_\xi \delta f = \int_{\partial\xi} f.$$

Next, Stokes' theorem tells us that for $\omega \in \mathscr{C}^1(U)$, $\phi \in C_2^s(U)$

$$\int_\phi \delta\omega = \int_{\partial\phi} \omega.$$

Finally, Gauss' theorem states that for $\alpha \in \mathscr{C}^2(U)$, $\tau \in C_3^s(U)$,

$$\int_\tau \delta\alpha = \int_{\partial\tau} \alpha.$$

In each case, as for Green's theorem above, it is enough to establish the result in a standard case (I, or T, or a standard tetrahedron), and study behavior under maps.

Arguing as before, we obtain maps

$$\mathscr{H}^0(U) \to \operatorname{Hom}\big(H_0(U), \mathbf{R}\big),$$
$$\mathscr{H}^1(U) \to \operatorname{Hom}\big(H_1(U), \mathbf{R}\big),$$
$$\mathscr{H}^2(U) \to \operatorname{Hom}\big(H_2(U), \mathbf{R}\big),$$

and it can be shown that all three of these are isomorphisms. We will not give the proof here.

THE COMPLEX CASE

Our object here again is not to give proofs of new, deep results so much as to show how standard theorems about one complex variable fit into the present framework. We return to $\mathbf{R}^2 = \mathbf{C}$ and write $z = x + iy$.

Consider a complex 1-form

$$\begin{aligned}\omega &= f(z)\,dz\\ &= \big(a(x, y) + ib(x, y)\big)(dx + i\,dy)\\ &= (a\,dx - b\,dy) + i(b\,dx + a\,dy).\end{aligned}$$

Then we have

$$\delta\omega = \left\{-\left(\frac{\partial a}{\partial y} + \frac{\partial b}{\partial x}\right) + i\left(\frac{\partial a}{\partial x} - \frac{\partial b}{\partial y}\right)\right\} dx\,dy$$

and so, in particular, $\delta\omega = 0$ if and only if

$$\frac{\partial b}{\partial y} = \frac{\partial a}{\partial x}, \qquad \frac{\partial b}{\partial x} = -\frac{\partial a}{\partial y}.$$

We recognize these as the Cauchy-Riemann equations, which express the fact that $f = a + ib$ is a differentiable function of z. Thus for f differentiable on an open set $U \subset \mathbf{C}$ and any $\phi \in C_2^s(U)$, we have

$$\int_{\partial\phi} \omega = \int_{\phi} \delta\omega = 0$$

by the above, on taking real and imaginary parts. This, of course, is a form of Cauchy's theorem.

Proceeding as before, we are led to define a group $\mathscr{H}^1_{\mathbf{C}}(U)$, the quotient of the group of 1-forms $\omega = f(z)\,dz$ with f differentiable on U by the subgroup consisting of the $g'(z)\,dz$, with g differentiable on U. (Of course, in the complex case, this already implies infinite differentiability.) Using line integrals, we obtain a map

$$J: \mathscr{H}^1_{\mathbf{C}}(U) \to \operatorname{Hom}\big(H_1(U), \mathbf{C}\big),$$

which can again be shown to be an isomorphism.

If U is the complement in $\mathbf{C}$ of a compact set K, the duality theorem of the preceding chapter gives an isomorphism

$$I_2: H_1(U) = H_1(\mathbf{C} - K) \to H^0(K).$$

This can be seen particularly clearly if K is a finite set of points $\{P_1, \ldots, P_N\}$, for then $H^0(K)$ is the free abelian group of maps $f: K \to \mathbf{Z}$, generated by the maps e_r with

$$\begin{cases} e_r(P_r) = 1, & \\ e_r(P_s) = 0 & (r \neq s); \end{cases}$$

and an element α of $\operatorname{Hom}\big(H^0(K), \mathbf{C}\big)$ is specified by the N complex numbers $\alpha_r = \alpha(e_r)$. Now given $\omega = f\,dz$, with f holomorphic on U, defining an element of $\mathscr{H}^1_{\mathbf{C}}(U)$, we can compute $\alpha = J(f)$ as follows.

For each r, $1 \leqslant r \leqslant N$, we have $e_r \in H^0(K)$. The element $I_2^{-1}(e_r)$ of $H_1(U)$ is represented by a small loop

$$l_r: z = P_r + \varepsilon\, e^{2\pi i t} \qquad (0 \leqslant t \leqslant 1)$$

encircling P_r once. (See Exercise 14.5.) Then

$$\alpha_r = \int_{l_r} \omega = \oint f(z)\,dz,$$

and of course f has an isolated singularity at P_r, and we have

$$\alpha_r = 2\pi i \operatorname{res}_{P_r} f$$

where $\operatorname{res}_{P_r} f$ denotes the residue of f at P_r.

FURTHER DEVELOPMENTS

Apart from the obvious generalization to n dimensions of the above, the next step is to formulate the results for manifolds, not just for open subsets of euclidean space. A convenient reference is Spivak. This gives a general form of Stokes' theorem, and so one can define a map generalizing Theorem 15.3. That this is an isomorphism is de Rham's theorem; there are proofs in de Rham and Hodge, for example. The latter book gives further results on differential forms which we cannot touch on here. For the general theory in the complex case, see Weil.

Hodge, W. V. D., *The Theory and Applications of Harmonic Integrals*, Cambridge University Press, 1941 (2nd edn. 1952).

de Rham, G., *Variétés Différentiables*, Hermann, Paris, 1955.

Spivak, M., *Calculus on Manifolds*, Benjamin, New York, 1965.

Weil, A., *Introduction à l'étude des Variétés Kählériennes*, Hermann, Paris, 1958.

EXERCISES AND PROBLEMS

1. Let U be obtained from $U(0, 2)$ in $\mathbf{R}^2$ by removing the points $(\pm 1, 0)$ and $(0, \pm 1)$. Show that Theorem 15.3 is surjective by constructing appropriate 1-forms.

2. For 1-forms

$$\phi = a_1\,dx_1 + a_2\,dx_2 + a_3\,dx_3,$$

$$\omega = b_1\,dx_1 + b_2\,dx_2 + b_3\,dx_3,$$

define

$$\phi \wedge \omega = (a_2 b_3 - a_3 b_2)\,dx_2\,dx_3 + (a_3 b_1 - a_1 b_3)\,dx_3\,dx_1 + (a_1 b_2 - a_2 b_1)\,dx_1\,dx_2.$$

Evaluate $\delta(\phi \wedge \omega)$, and interpret in terms of the vector calculus.

3. Let U be the unit ball $U(0, 1)$ in $\mathbf{R}^3$. Show that the $\mathscr{H}^i(U) \to \operatorname{Hom}\big(H_{2-i}(U), \mathbf{R}\big)$ described in the text are isomorphisms, by generalizing the last part of Theorem 15.3. (This is the first step in the proof of the general case.)

INDEX OF TERMS

INDEX OF NOTATION

A CATALOG OF SELECTED

DOVER BOOKS

IN SCIENCE AND MATHEMATICS

RELATIVITY, THERMODYNAMICS AND COSMOLOGY, Richard C. Tolman. Landmark study extends thermodynamics to special, general relativity; also applications of relativistic mechanics, thermodynamics to cosmological models. 501pp. 5⅜ × 8½. 65383-8 Pa. $12.95

APPLIED ANALYSIS, Cornelius Lanczos. Classic work on analysis and design of finite processes for approximating solution of analytical problems. Algebraic equations, matrices, harmonic analysis, quadrature methods, much more. 559pp. 5⅜ × 8½. 65656-X Pa. $12.95

SPECIAL RELATIVITY FOR PHYSICISTS, G. Stephenson and C.W. Kilmister. Concise elegant account for nonspecialists. Lorentz transformation, optical and dynamical applications, more. Bibliography. 108pp. 5⅜ × 8½. 65519-9 Pa. $4.95

INTRODUCTION TO ANALYSIS, Maxwell Rosenlicht. Unusually clear, accessible coverage of set theory, real number system, metric spaces, continuous functions, Riemann integration, multiple integrals, more. Wide range of problems. Undergraduate level. Bibliography. 254pp. 5⅜ × 8½. 65038-3 Pa. $7.95

INTRODUCTION TO QUANTUM MECHANICS With Applications to Chemistry, Linus Pauling & E. Bright Wilson, Jr. Classic undergraduate text by Nobel Prize winner applies quantum mechanics to chemical and physical problems. Numerous tables and figures enhance the text. Chapter bibliographies. Appendices. Index. 468pp. 5⅜ × 8½. 64871-0 Pa. $11.95

ASYMPTOTIC EXPANSIONS OF INTEGRALS, Norman Bleistein & Richard A. Handelsman. Best introduction to important field with applications in a variety of scientific disciplines. New preface. Problems. Diagrams. Tables. Bibliography. Index. 448pp. 5⅜ × 8½. 65082-0 Pa. $12.95

MATHEMATICS APPLIED TO CONTINUUM MECHANICS, Lee A. Segel. Analyzes models of fluid flow and solid deformation. For upper-level math, science and engineering students. 608pp. 5⅜ × 8½. 65369-2 Pa. $13.95

ELEMENTS OF REAL ANALYSIS, David A. Sprecher. Classic text covers fundamental concepts, real number system, point sets, functions of a real variable, Fourier series, much more. Over 500 exercises. 352pp. 5⅜ × 8½. 65385-4 Pa. $10.95

PHYSICAL PRINCIPLES OF THE QUANTUM THEORY, Werner Heisenberg. Nobel Laureate discusses quantum theory, uncertainty, wave mechanics, work of Dirac, Schroedinger, Compton, Wilson, Einstein, etc. 184pp. 5⅜ × 8½. 60113-7 Pa. $5.95

INTRODUCTORY REAL ANALYSIS, A.N. Kolmogorov, S.V. Fomin. Translated by Richard A. Silverman. Self-contained, evenly paced introduction to real and functional analysis. Some 350 problems. 403pp. 5⅜ × 8½. 61226-0 Pa. $9.95

PROBLEMS AND SOLUTIONS IN QUANTUM CHEMISTRY AND PHYSICS, Charles S. Johnson, Jr. and Lee G. Pedersen. Unusually varied problems, detailed solutions in coverage of quantum mechanics, wave mechanics, angular momentum, molecular spectroscopy, scattering theory, more. 280 problems plus 139 supplementary exercises. 430pp. 6½ × 9¼. 65236-X Pa. $12.95

ROTARY-WING AERODYNAMICS, W.Z. Stepniewski. Clear, concise text covers aerodynamic phenomena of the rotor and offers guidelines for helicopter performance evaluation. Originally prepared for NASA. 537 figures. 640pp. 6⅛ × 9¼.
64647-5 Pa. $15.95

DIFFERENTIAL GEOMETRY, Heinrich W. Guggenheimer. Local differential geometry as an application of advanced calculus and linear algebra. Curvature, transformation groups, surfaces, more. Exercises. 62 figures. 378pp. 5⅜ × 8½.
63433-7 Pa. $8.95

INTRODUCTION TO SPACE DYNAMICS, William Tyrrell Thomson. Comprehensive, classic introduction to space-flight engineering for advanced undergraduate and graduate students. Includes vector algebra, kinematics, transformation of coordinates. Bibliography. Index. 352pp. 5⅜ × 8½. 65113-4 Pa. $8.95

A SURVEY OF MINIMAL SURFACES, Robert Osserman. Up-to-date, in-depth discussion of the field for advanced students. Corrected and enlarged edition covers new developments. Includes numerous problems. 192pp. 5⅜ × 8½.
64998-9 Pa. $8.95

ANALYTICAL MECHANICS OF GEARS, Earle Buckingham. Indispensable reference for modern gear manufacture covers conjugate gear-tooth action, gear-tooth profiles of various gears, many other topics. 263 figures. 102 tables. 546pp. 5⅜ × 8½. 65712-4 Pa. $14.95

SET THEORY AND LOGIC, Robert R. Stoll. Lucid introduction to unified theory of mathematical concepts. Set theory and logic seen as tools for conceptual understanding of real number system. 496pp. 5⅜ × 8¼. 63829-4 Pa. $10.95

A HISTORY OF MECHANICS, René Dugas. Monumental study of mechanical principles from antiquity to quantum mechanics. Contributions of ancient Greeks, Galileo, Leonardo, Kepler, Lagrange, many others. 671pp. 5⅜ × 8½.
65632-2 Pa. $14.95

FAMOUS PROBLEMS OF GEOMETRY AND HOW TO SOLVE THEM, Benjamin Bold. Squaring the circle, trisecting the angle, duplicating the cube: learn their history, why they are impossible to solve, then solve them yourself. 128pp. 5⅜ × 8½. 24297-8 Pa. $4.95

MECHANICAL VIBRATIONS, J.P. Den Hartog. Classic textbook offers lucid explanations and illustrative models, applying theories of vibrations to a variety of practical industrial engineering problems. Numerous figures. 233 problems, solutions. Appendix. Index. Preface. 436pp. 5⅜ × 8½. 64785-4 Pa. $10.95

CURVATURE AND HOMOLOGY, Samuel I. Goldberg. Thorough treatment of specialized branch of differential geometry. Covers Riemannian manifolds, topology of differentiable manifolds, compact Lie groups, other topics. Exercises. 315pp. 5⅜ × 8½. 64314-X Pa. $8.95

HISTORY OF STRENGTH OF MATERIALS, Stephen P. Timoshenko. Excellent historical survey of the strength of materials with many references to the theories of elasticity and structure. 245 figures. 452pp. 5⅜ × 8½. 61187-6 Pa. $11.95

GEOMETRY OF COMPLEX NUMBERS, Hans Schwerdtfeger. Illuminating, widely praised book on analytic geometry of circles, the Moebius transformation, and two-dimensional non-Euclidean geometries. 200pp. 5⅜ × 8¼.
63830-8 Pa. $8.95

MECHANICS, J.P. Den Hartog. A classic introductory text or refresher. Hundreds of applications and design problems illuminate fundamentals of trusses, loaded beams and cables, etc. 334 answered problems. 462pp. 5⅜ × 8½. 60754-2 Pa. $9.95

TOPOLOGY, John G. Hocking and Gail S. Young. Superb one-year course in classical topology. Topological spaces and functions, point-set topology, much more. Examples and problems. Bibliography. Index. 384pp. 5⅜ × 8¼.
65676-4 Pa. $9.95

STRENGTH OF MATERIALS, J.P. Den Hartog. Full, clear treatment of basic material (tension, torsion, bending, etc.) plus advanced material on engineering methods, applications. 350 answered problems. 323pp. 5⅜ × 8½. 60755-0 Pa. $8.95

ELEMENTARY CONCEPTS OF TOPOLOGY, Paul Alexandroff. Elegant, intuitive approach to topology from set-theoretic topology to Betti groups; how concepts of topology are useful in math and physics. 25 figures. 57pp. 5⅜ × 8½.
60747-X Pa. $3.50

ADVANCED STRENGTH OF MATERIALS, J.P. Den Hartog. Superbly written advanced text covers torsion, rotating disks, membrane stresses in shells, much more. Many problems and answers. 388pp. 5⅜ × 8½. 65407-9 Pa. $9.95

COMPUTABILITY AND UNSOLVABILITY, Martin Davis. Classic graduate-level introduction to theory of computability, usually referred to as theory of recurrent functions. New preface and appendix. 288pp. 5⅜ × 8½. 61471-9 Pa. $7.95

GENERAL CHEMISTRY, Linus Pauling. Revised 3rd edition of classic first-year text by Nobel laureate. Atomic and molecular structure, quantum mechanics, statistical mechanics, thermodynamics correlated with descriptive chemistry. Problems. 992pp. 5⅜ × 8½. 65622-5 Pa. $19.95

AN INTRODUCTION TO MATRICES, SETS AND GROUPS FOR SCIENCE STUDENTS, G. Stephenson. Concise, readable text introduces sets, groups, and most importantly, matrices to undergraduate students of physics, chemistry, and engineering. Problems. 164pp. 5⅜ × 8½. 65077-4 Pa. $6.95

THE HISTORICAL BACKGROUND OF CHEMISTRY, Henry M. Leicester. Evolution of ideas, not individual biography. Concentrates on formulation of a coherent set of chemical laws. 260pp. 5⅜ × 8½. 61053-5 Pa. $6.95

THE PHILOSOPHY OF MATHEMATICS: An Introductory Essay, Stephan Körner. Surveys the views of Plato, Aristotle, Leibniz & Kant concerning propositions and theories of applied and pure mathematics. Introduction. Two appendices. Index. 198pp. 5⅜ × 8½. 25048-2 Pa. $7.95

THE DEVELOPMENT OF MODERN CHEMISTRY, Aaron J. Ihde. Authoritative history of chemistry from ancient Greek theory to 20th-century innovation. Covers major chemists and their discoveries. 209 illustrations. 14 tables. Bibliographies. Indices. Appendices. 851pp. 5⅜ × 8½. 64235-6 Pa. $18.95

THE FOUR-COLOR PROBLEM: Assaults and Conquest, Thomas L. Saaty and Paul G. Kainen. Engrossing, comprehensive account of the century-old combinatorial topological problem, its history and solution. Bibliographies. Index. 110 figures. 228pp. 5⅜ × 8½. 65092-8 Pa. $6.95

CATALYSIS IN CHEMISTRY AND ENZYMOLOGY, William P. Jencks. Exceptionally clear coverage of mechanisms for catalysis, forces in aqueous solution, carbonyl- and acyl-group reactions, practical kinetics, more. 864pp. 5⅜ × 8½. 65460-5 Pa. $19.95

PROBABILITY: An Introduction, Samuel Goldberg. Excellent basic text covers set theory, probability theory for finite sample spaces, binomial theorem, much more. 360 problems. Bibliographies. 322pp. 5⅜ × 8½. 65252-1 Pa. $8.95

LIGHTNING, Martin A. Uman. Revised, updated edition of classic work on the physics of lightning. Phenomena, terminology, measurement, photography, spectroscopy, thunder, more. Reviews recent research. Bibliography. Indices. 320pp. 5⅜ × 8¼. 64575-4 Pa. $8.95

PROBABILITY THEORY: A Concise Course, Y.A. Rozanov. Highly readable, self-contained introduction covers combination of events, dependent events, Bernoulli trials, etc. Translation by Richard Silverman. 148pp. 5⅜ × 8¼. 63544-9 Pa. $5.95

AN INTRODUCTION TO HAMILTONIAN OPTICS, H. A. Buchdahl. Detailed account of the Hamiltonian treatment of aberration theory in geometrical optics. Many classes of optical systems defined in terms of the symmetries they possess. Problems with detailed solutions. 1970 edition. xv + 360pp. 5⅜ × 8½. 67597-1 Pa. $10.95

STATISTICS MANUAL, Edwin L. Crow, et al. Comprehensive, practical collection of classical and modern methods prepared by U.S. Naval Ordnance Test Station. Stress on use. Basics of statistics assumed. 288pp. 5⅜ × 8½. 60599-X Pa. $6.95

DICTIONARY/OUTLINE OF BASIC STATISTICS, John E. Freund and Frank J. Williams. A clear concise dictionary of over 1,000 statistical terms and an outline of statistical formulas covering probability, nonparametric tests, much more. 208pp. 5⅜ × 8½. 66796-0 Pa. $6.95

STATISTICAL METHOD FROM THE VIEWPOINT OF QUALITY CONTROL, Walter A. Shewhart. Important text explains regulation of variables, uses of statistical control to achieve quality control in industry, agriculture, other areas. 192pp. 5⅜ × 8½. 65232-7 Pa. $7.95

THE INTERPRETATION OF GEOLOGICAL PHASE DIAGRAMS, Ernest G. Ehlers. Clear, concise text emphasizes diagrams of systems under fluid or containing pressure; also coverage of complex binary systems, hydrothermal melting, more. 288pp. 6½ × 9¼. 65389-7 Pa. $10.95

STATISTICAL ADJUSTMENT OF DATA, W. Edwards Deming. Introduction to basic concepts of statistics, curve fitting, least squares solution, conditions without parameter, conditions containing parameters. 26 exercises worked out. 271pp. 5⅜ × 8½. 64685-8 Pa. $8.95